Improving Urban Science Education

Reverberations
Contemporary Curriculum and Pedagogy

Series Editors

Joe L. Kincheloe and Shirley R. Steinberg

Titles in the Series

Improving Urban Science Education

New Roles for Teachers, Students, and Researchers

EDITED BY KENNETH TOBIN, ROWHEA ELMESKY, AND GALE SEILER

ROWMAN & LITTLEFIELD PUBLISHERS, INC.
Lanham • Boulder • New York • Toronto • Oxford

ROWMAN & LITTLEFIELD PUBLISHERS, INC.

Published in the United States of America
by Rowman & Littlefield Publishers, Inc.
A wholly owned subsidiary of The Rowman & Littlefield Publishing Group, Inc.
4501 Forbes Boulevard, Suite 200, Lanham, MD 20706
www.rowmanlittlefield.com

P.O. Box 317, Oxford OX2 9RU, UK

British Library Cataloguing in Publication Information Available

Library of Congress Cataloging-in-Publication Data Available

ISBN 0-7425-3704-8 (cloth : alk. paper)
ISBN 0-7425-3705-6 (pbk. : alk. paper)

Printed in the United States of America

♾™ The paper used in this publication meets the minimum requirements of American
National Standard for Information Sciences—Permanence of Paper for Printed Library
Materials, ANSI/NISO Z39.48-1992.

Contents

Foreword

Student Voices on Urban Science Education: A Collage of Student Researcher Perspectives

Student Researchers

This foreword represents a patchwork of perspectives on urban science education and the role of students as researchers therein, woven together with the voices of the student researchers across research sites within our study. The foreword begins by embracing the traditionally silenced voices of urban youth as they speak to the inequities of urban schooling and issues of poor science teaching and learning, some of which are discussed on a macro level (school-wide structural constraints) and, in other sections, on an individual, microscopic level (students' personal school issues). Following the delineation of the challenges facing urban students, in the second portion of the foreword, student researchers turn their voices away from the deficits of the system to embrace a tone of empowerment. The result, we hope, is a powerful expression of how being a student researcher has impacted their lives, as well as how they impact the research process, thereby capturing the impact that student researchers, in general, can have on furthering our understandings of urban science education. While we, the editors, have arranged the writings of these youth into the structure of a foreword, it is their agency as student researchers which in many ways *structures* how this foreword unfolds, just as they in many ways *structured* how our research has unfolded, and the powerful meanings and perspectives that have been gained.

The Experience of Urban Schooling: A Student Perspective

The First Step: Staying in School (*Tyronne*)

Students go to school for lots of reasons, some 'cause they have to, some 'cause they know they need to. You can find lots of students that come to school to further their education, but you can also find lots of students who try to stop other students from reaching their goals. I think if the students feel more motivated they would not mind coming to school every day and doing what they are supposed to do.

I think the biggest reason students don't like to come to school is because of what you call a suspension. "Why come to school when you're going to get put right back out?" That's what most students say. Many students get suspended every day at City High, for things as small as being in the hallways. I think that's wrong to put a student out of school, stopping them from getting their education, when the people who work at the school already have theirs. Why tell a child how important it is to have and achieve something, and then stop them from entering the door to get it?

Many students handle problems at home, so why do they have come to school and handle more problems? You are told to leave your problems at home —that's what they tell the students. But some NTAs (nonteaching assistants) bring their problems to school and take it out on the students. The students should be respected as well as the NTAs are getting respected, then the school day can flow easier.

When I first came to City High I was placed in *Business* small learning community, and I had no idea what I was getting into. I tried to start the year off good, but it didn't work out that way. I started getting into trouble with NTAs so much I didn't know what was happening. When I was in the hall, with or without permission, I seemed to get suspended every time. I was starting to think they didn't want me there, because they put me out so much. I missed so many days that year, I was held back from moving to the tenth grade.

The next year I thought it might be a little different, when they placed me in *Incentive*, but it was still the same. That year I was told I could not for any reason come on the second floor. Sometimes when I would come from lunch (which was on the second floor) I would get in trouble for being up there. I would try to tell them why I was there, but it didn't work. That year, I had to repeat the ninth grade again.

My third year, when I should have been in the eleventh grade, I was in the same grade as my little sister—the ninth. That year seemed a little better; I was in school more. My grades shot up higher than I had seen in a long time. My guess is that this happened because I had seen more of the inside of the school that year than I had seen in the other two years I had been at City High.

So, Now That I'm Here: What Happens in Science Class (*Natalie*)

As the years go by it seems that I have done worst in the sciences, the reasons being: my attention span and tolerance for being idle. When I say "my attention span," I mean that when I get bored I acquire a short attention span. When this occurs I tend to sleep, disrupt the class, talk to the other students, and other things of that nature. Therefore, my work doesn't get completed. When I say my "tolerance for being idle," I mean my zero tolerance for having nothing to do. This only happens when I finish my work before the rest of the class and then I have nothing to do. I absolutely dislike having nothing to do. In that case, by finishing early, my work sometimes gets misplaced or destroyed, and very seldom, turned in. Actually, the only work that does get turned in is the work that is turned in as soon as it is completed. The other work gets lost because it is mixed up in all of the junk that I have and which, by this point, I have taken out to try to find something else to occupy myself with. These actions, in the long run, cause my grade to fall to a near failing grade. Not the best reasons, but the reasons, nonetheless.

I am not bored in all of my science classes, just chemistry and physics. Chemistry bores me because I was enrolled in it after I had taken biochemistry, which is much more in depth than chemistry. To me, chemistry is more book-work than anything, whereas biochemistry is more lab work to me, and it was fun. Physics is boring to me because all it is just a group of measurements that you will never use again in life. Physics is one of these classes that you really don't need, but you have to take it because they (the school administration) say so. The only two challenges that I had in my physics class were drawing the pictures for the magnitude and displacement and staying awake in class.

In my science class no one helps me other than to tell me to stay awake in class. But my friends help me to stay on task as far as homework, now. Other than the usual, there's nothing much to say about homework. We don't really get much of it, but we do get homework, and sometimes it's useful and sometimes it's not.

Becoming a Student Researcher

This was no ordinary job. It was like getting paid the ultimate summer school.
Getting paid to learn is off the hook. Not only did I learn science, I learned that
most people will help you if you know how to ask. Last summer was like a link,
no a chain, you never know when it's going to end. Being around people who
care is a great feeling. When I was down, needed help or someone to talk to,
my work family was here. (Shakeem, journal, 8/02)

What Does Being a Student Researcher Really Mean? (*Jessica*)

What exactly is a student researcher? Well, according to Webster's a student is
"a person formally engaged in learning." Webster's also defines a researcher as
"one who makes an extensive investigation into discovering or revising facts and
theories." When you put the two definitions together you get a job that most people will think is boring. Well, I must admit I thought the job was dull at first, too.
At first I only took the job because we received free pizza, cheese fries, and soda
every time we had a meeting. I figured if I came to these meetings I could get
free food and all I have to do is tell people how I feel about science. Then when I
really became involved in the research I realized that being a student researcher
is not that bad and that it has some good things about it.

Getting Recruited (*Ivory*)

Hi, my name is Ivory. Boy, I tell you if it wasn't for me attending City High and
running into some of the kindest people, I probably would have had it hard. I
would not have met people that care so much about me and my education. It
wasn't the school that changed me, it was the people in the school. It all began
when I was in the ninth grade, at the end of the school year when my teacher,
Ms. Jen Beers, called me on the phone and asked if I'd like to become a student
researcher. She told me it was a project that Dr. Tobin at the University of Pennsylvania was working on and he needed some students to help out. I asked what
type of project it was and how long it would be. She said it was a grant funded
by the National Science Foundation (NSF) to study how to improve the teaching
and learning of science in urban classrooms. Being as though I needed to improve my science, as well as many other subjects, I said yes and accepted the job
of becoming a student researcher. So, I've been a researcher since the summer of
2001 and worked for the summers of 2002 and 2003 as well.

In the beginning, I really didn't know what I was up against. I wasn't expecting to be working at a college in a science education department helping teachers and learning at the same time, but things started to fall in place. I guess it was sort of interesting being the teacher educator for once—having them learn from us. My role as a student researcher wasn't easy at first. We did a lot of work and learning. The hardest lesson was remembering we were no longer the students but we were now the teacher, which meant we had to make our own lessons and develop a new curriculum. It was hard for us to get used to that for a while. But as time went on, things changed.

Kids and Adults Can Work on the Same Level (*Shakeem*)

I didn't know I'd actually be learning, and doing video editing, until I got here. It just happened. I said OK. I had an opinion and they said OK, we're gonna listen to you. So I gave them the opinion. They listened and it worked or they listened and it didn't work, whatever the situation was. It was all on the same level. In high school there's a lot of people who are, let's say, above you and it's like here we're all equals. Of course I'm not an adult but if I have an opinion, it's heard. At school if you tell your teacher I got an opinion, most likely they'll say OK we're gonna get to it, but most likely it never gets done. So I think that since everybody is more on the same level, everyone listens here and in school no.

It should happen (this way). It should happen a lot more. So I don't know why it seems strange that kids and adults can work on the same level. I think it should be a lot more. We were pretty much in charge here all the time. It's like we were in charge and they were just basically saying you've got the power and we want you to come up with the full potential that you have.

Student Researcher Roles: A Sampling

Creation of Curriculum Resources (*May*)

The university researchers wanted us to create a new curriculum for teaching the physics of sound since they thought it would be easier for students to learn if it came from other students' perspectives. That was a hard topic and a big challenge. Since we were used to being the students we had no clue how to come up with lesson plans or how it would work. To do this we had to be both the teacher and the student and that felt kind of awkward. They wanted us to begin by doing Internet research to find websites that provided information on sound-related concepts, such as sound as mechanical vibrations, sound as the movement of energy in the form of waves, wavelength for transverse and longitudinal type waves, and physical interpretations of frequency as pitch. So we began doing what they asked but things weren't going so right, so we figured we'd take matters into our own hands.

What we added. We sat around for awhile thinking to ourselves and sharing ideas on how to make this project complete but still we had no clue. After a couple days of hard thinking we started to give up and just do what the researchers suggested, until someone said let's make a movie. I thought it could never happen, and I thought that it was just a waste of time but the others seemed real into it so I agreed. We started by watching a show called Bill Nye the Science Guy. Everyone laughed at the show we used to love and said we could do better. We also watched a variety of different movies to get ideas, like "Men of Honor" and "The Right Stuff." We also did a lot of research on the Internet about sound waves, to learn how sound could harm you and also how it could help you. I learned about a couple of different types of waves and about the concepts of pitch and frequency. Dr. Tobin also took us on a walk through the city locating different sounds and getting different examples for our movie.

As part of our movie, we all wanted to do skits on things we liked doing or wanted to do in the future. Since I want to be a massage therapist, I research things on sound about that, and actually found something! I found that masseuses use soft music to ease their patients' minds. As a result, I created a skit to communicate this information. I think this skit was important to add in the movie because it gave me, as a student, a different perspective of sound and on how important it was to people, other then us. Another girl named Ivory did a skit on how sound can break this little bone in your ear. She explained, in detail, about how you have three little bones in your ear and how if sound gets too loud, they could break it. Shakeem, one of the male student researchers, did a rap with Ivory, and I wrote some poems. Two of the other males, Kareem and Randy, worked on poster boards. After we videotaped everyone's skits, we had to do video editing. This was a long process since we only had one machine, but it worked out OK.

In the movie we made, "Sound in the City," we tried to teach kids about sound waves, amplitude, frequency, and many other concepts. This is one of the biggest projects I have ever been involved with, and I hope you enjoy reading about it because I feel that it was a big success.

How Being a Student Researcher Has Changed Me
(*Jessica and Ivory*)

Being a student researcher has helped me develop skills in areas that I would never have had the opportunity to develop any other place. While being a student researcher, I have learned to use computer programs—iMovie and Adobe Premiere. Learning to use these programs has proven to be useful when I need to do a class project or just to make a simple movie for myself. I also learned how to do a PowerPoint presentation. Learning to use PowerPoint was the best thing I could have done because my sister had to do a PowerPoint presentation and she was completely confused. I had to show her how to do one and I am eight years younger then her.

Being a student researcher not only helps you look at things from another perspective but it also helped me to look past the things that are obvious. Being a student researcher has its own special way of causing you to think. A simple discussion about a day at school could turn into a complex thinking question such as why do you think people act the way they do? With student research there is an endless trail of questions. One question always leads to another and you can go so far you forget what the conversation was originally about.

Being a student researcher is also not a one man's job. I work with three other people and we laugh all day, but when it comes down to work we calm down a little bit or at least long enough to answer some questions. We feed off of each other's energy as well as ideas. We have debates about whose idea is right and why. We just have so much fun in those two or three hours we are together. I went to school with the three people I work with but being student researchers and working together has showed me another side of them. I see how different our thought patterns are, and I see how our views sometimes contradict each other's. The contradicting views bring up something else that being a student researcher has taught me. It taught me how to respect another person's opinion. When I am at home, I will just straight out tell you that your opinion stinks. Being a student researcher helped me to just listen to people's opinions and if I disagree with them, I state my opinion and explain why I like it better than the other opinion.

As time went on I began to like being a student researcher, and I was no longer in it for the food. I was in it because I was interested in the things that being a student researcher had to offer. Being a student researcher equips you with skills you can use beyond the field of science and, when I really think about it, I am glad I took the opportunity to become a student researcher. Hopefully the skills I learn from being a student researcher will remain with me always. (*Jessica*)

> DUS is back.
> Second time around so you know we can't be wacked.
> Four kids teachin' science, who messin' with that?
> Thought on the last tape that we toldja
> You messin' with a brand new line of researchers
> My team is nice and yours is good.
> But we got the best and brightest from the worst of hoods
> DUS stand on top above the rest
> So I guess the best that we can claim
> Puttin' Bill Nye the Science Guy and Hank to shame.
> Degrees on the wall that read our name
> Ivory, Shakeem, May and R, killin' the game.
> (Rap presentation, "Sickin," Ivory and Shakeem, Summer 2002)

My experience being a student researcher was interesting. It was exciting for me because I got to show my sports and other talents, which are part of my life, like rapping and basketball, by using what I know as learning sources. The impact that DUS put on my life was to be a hard-working student. I also learned that when answering a question, always give reasons for why you are choosing it. I can say that the years I spent being a researcher, I learned a lot in a short time. I know how to video edit movies for example. I also learned how to transcribe audiotapes. My favorite was video editing in which I used to make rap videos as well as movies off of the iMovie program and transfer them to QuickTime. Since I've been a part of DUS, I have finished three rap videos, which were called Gender, DUS, and Sickin. During my last summer with DUS, we worked on lesson planning for high school students. Our job was to make lessons for the students more interesting and focused. I can say that these three years with the group have been very exciting and fun years in my life. This is something I can look back at years from now and say to myself, and explain to younger kids, that I've changed. I appreciate all of the teachers that were part of the program, as well as my life. (*Ivory*)

Looking Forward from the Foreword

There are numerous ways in which a book on urban science education could begin. Perhaps a strong case could be made for a foreword written by hard-working teachers, powerful administrators, prestigious scholars, or university researchers. Yet, by beginning this book with a foreword containing the voices of student researchers, we make a statement about the essential need for discussions around urban science education to not only include, but to also privilege, the students attending this nation's inner city schools. Their insights, feelings, perspectives, and contributions as student researchers have been an irreplaceable part of our research process and will continue to appear throughout the book, in various chapters, to add dimension and depth to the studies undertaken across the school sites.

Although as a population, we may share the common experience of "schooling" with our student researchers, their lived experiences within the structures of schools and classrooms differ in dynamic ways from those of us who come from the mainstream dominant culture. With their assistance, we are able to develop understandings of how structure constrains and affords their agency. Moreover, we have been become more able to recognize examples of students' agency within and across fields, inside and outside of school, which may look "different," since they may be utilizing resources and adhering to schema that are unfamiliar to those with differing sociocultural histories.

> Now there no stoppin' us now
> Soon we gonna get paid
> Like the Planet of the Apes
> Ridin' in 4 by 4 Drango
> Ice everywhere
> They're even on Dingos
> DUS Dog tags
> How that sound, Kid?
> *The world waitin'.*
> The name EB Marvelous.
> Got no time 4 ya'll debatin'
> (Rap presentation, "Sickin," Ivory and Shakeem, Summer 2002)

In this nation, where oppression is insufferable and individuals are encouraged to be the best they can be, a move to a more equitable urban schooling experience is more readily mobilized by changing the structure of how we conduct educational research. "The land of the free," we reiterate again and again as we teach our students American history. Yet in actuality, freedom, or an actualization of one's own agency, can never be obtained through silencing.

Reflected in the above lyrics, Ivory's (EB Marvelous) confidence, anticipation of success in life and her strong identification with a "squad" (Discovering Urban Science) formed around a commitment to research and science captures precisely the empowerment we advocate for urban youth in compiling a book focused on the reforming of urban science education. As Ivory so explicitly conveys through three simple words, "the world waitin'," the reformation of science education in urban settings can not occur successfully if we only expect the students to change. "The world waitin'" implies that youth like Ivory are ready to participate within society in powerful ways. The question is: Is the world ready for our urban youth and the multitude of rich resources they embody?

In this book, we embrace Ivory's challenge to the world and agree that we have debated too long about what urban youth do not know, do not want to know, will not do, or cannot do *rather than* what they do know, want to know and will, can, and should do (Boykin, 1986). We join hands in solidarity with Ivory and all of the students and student researchers whom have touched our lives. We are committed to affording their voices, not only in a concrete sense by including their artifacts, but more substantially, by advocating the restructuring of schools and schooling to respond to their perspectives and to promote their sense of agency.

Acknowledgments

The research in this book is supported in part by the National Science Foundation under Grant No. REC-0107022. Any opinions, findings, and conclusions or recommendations expressed in the book are those of the authors and do not necessarily reflect the views of the National Science Foundation.

In any project of this magnitude there are many stakeholders and participants who we want to thank for their active support. We have been astonished at the generosity and awesome talents of the students and very many of them have assisted us to better understand their lives and have willingly participated intensively and extensively in our research in myriad ways. A special category of students is the student researcher. Across the participating schools we employed and otherwise engaged a great many student researchers, far too many to single out by naming them here. We are deeply indebted to the work, optimism, and tenacity of the urban youth with whom we have an enduring bond.

A smaller number of teachers have been integrally involved in our research. We extend our deep appreciation to these teachers and assure them that we have the utmost respect for their professional commitments and the magnitude of the challenge and opportunity of teaching in urban schools.

A handful of administrators also have been of enormous assistance to us in undertaking this research and many of them have provided data resources, participated in the analysis and interpretation of data, planned for further data gathering, and read our research and enacted those parts that had the potential to improve the quality of science education in their schools. Their involvement went beyond just providing access to include full participation in learning from the research and ensuring that we learned as much as possible from our efforts.

We have had our own squad and it has operated in a scholarly way for the duration of the research. Old-timers such as Gale have been only too willing to share their knowledge with newcomers, such as Rowhea when she first arrived. And then, Rowhea's willingness to mentor other researchers, such as Stacy and Sarah-Kate was highly appreciated. As the research group has increased its level of expertise, the mentor group has expanded and it is quite apparent that we have a relatively large group of mentors who have made it possible for teacher researchers and student researchers to build deep understandings of the theoretical frameworks we have employed and to learn how to undertake research at the micro- and meso- levels of social life. The squad has been dynamic and as some people have moved to different universities, others have joined us. However, none left entirely and we are always learning from interactions with participants (electronic and face-to-face).

Finally, support from others is essential and, especially in the long months of editing the chapters that compose this book, the editors were all supported by family and loved ones. To those who are closest to us we express our deep gratitude.

Introduction

Kenneth Tobin, Rowhea Elmesky, and Gale Seiler

This volume is a systematic attempt to document what we have learned from seven years of research on the teaching and learning of science in the urban high schools of Philadelphia. The chapters in the book address issues that are salient to the major problems that currently face urban science education and include the perspectives of students (foreword), teachers (chapters 3, 7, 8, 9, 12, 13, 15), and university researchers (chapters 1, 2, 3, 4, 5, 6, 8, 10, 11, 12, 14, 16, 17). The following sections of this introduction contain a description of each of the chapters in the book.

Navigating the Book

The first chapter of the book takes seriously the goal of shattering deficit views of urban African American students as it reviews some critical aspects of education in Philadelphia and describes the methodology employed in the study.

In chapter 2 Kenneth Tobin explores his teaching of science at City High over a period of four years during which he endeavored to address the pervasive problems of teaching and learning science in urban neighborhood schools. After a lengthy period in which he was highly unsuccessful, he was able to teach in ways that adapted to the cultural resources that students brought from fields away from the classroom and create learning environments that were conducive to the students being actively involved. The chapter shows how Tobin's practices adapted over time and as they did so different ways of participating were evident in the students. The class began to flow and there was evidence of chains of successful interactions, synchrony, and the development of solidarity among students. Tobin shows that students were able to assume leadership roles in science classrooms, monitor potentially disruptive practices, and use practices from other fields in successful interactions that led to deeper insights into science. The study emphasizes the importance of the reciprocity of students and teachers learning about one another's culture and adapting their practices to be accommodating to the participation and successes of others.

The focus of chapter 3 is City High, the site of our longest research involvement. The coauthors (Seiler and Butler) employ a cultural-historical approach to examine the material and human resources present in the school, and how these interact in a ying and yang fashion that illustrates the ways in which agency and structure transform and constitute each other in this field.

In chapter 4 Wolff-Michael Roth examines the organization of schools in terms of agency | structure relationships and the contradictions that can arise when boundary objects emanating from one field cross into another. Roth's theorizing and analyses span the micro, meso, and macro levels and show how processes at one level structure practices at another. His focus on the temporal and spatial characteristics of fields and their coordination as organizations is supported with theoretical and empirical analyses. Finally, Roth also shows clearly how, within agency | structure relationships the individual also is dialectically interconnected with the collective.

Elmesky addresses the potential of cultural capital created in other fields as a basis for creating science fluency and communities in which solidarity can emerge and an associated willingness to participate in science-related activities. Through an exploration of *playin* she highlights the potential pitfalls associated with forms of teaching that are not cognizant of the culture of urban youth and which do not create resonant structures to afford flow in science classrooms. Chapter 5 emphasizes the positive aspects of what urban youth bring to the classroom, especially their capacities to multitask and coparticipate in fields in which oral fluency is combined with rhythm and verve. The chapter contains micro and meso analyses and the cross field analyses extend to critical insights into issues of agency and structure at the macro level as well.

In chapter 6 Seiler examines the practices of African American youth, especially males, through lenses that are explicitly nondeficit. She adopts the perspectives of cultural sociology and black psychology in analyses that emphasize the impressive oral traditions of youth and their ways of knowing that embrace analogies and practices that extend beyond what can be re-presented as oral and written texts. Her chapter emphasizes the potential for teaching and learning of science of identifying the cultural capital of youth and enacting curricula in ways that allow them to pursue their interests and use what they know and can do as a foundation for science fluency. A feature of the chapter is an analysis that spans micro, meso, and macro social levels and explores capital from other fields in the field of the science classroom.

Beers explores her journey of becoming an urban science teacher from her experiences as a student teacher at City High through her first two years as a beginning teacher at Charter High. Since she is also the teacher involved in chapter 6 there is continuity of chapters 6 and 7. Beers emphasizes the necessity to learn about the culture of urban youth, incorporate the interests of students into enacted curricula, and demonstrate trust of the students and a willingness to listen to their voices. The chapter provides clear examples of what urban students can accomplish in science education.

LaVan and Beers describe in chapter 8 how they used cogenerative dialogues, small group conversations that included them as coteachers and student researchers who were selected to be as different from one another as possible. The goal of the cogenerative dialogues is to create collective responsibilities for the quality of science education; including the roles to be enacted by participants and the associated division of labor, the rules that will apply to participation, what resources, including language, are accessible and who will use them at specific times, and what is to be accomplished collectively. Central to the use of cogenerative dialogues was the use of video tapes that were selected for their salience to patterns of coherence and associated contradictions that were then discussed. LaVan and Beers also applied cogenerative dialogues as transformative activities with the whole class.

In chapter 9 Carambo probes the centrality of including students' interests in the science curriculum. Through the use of activity theory and cultural sociology he explores what students can learn and accomplish when they are permitted to choose activities that involve the doing of science and adapt their goals as the activity proceeds. His study explores science fluency within an overall analysis of the agency | structure dialectic. The chapter has several points of intersection with earlier chapters, including the students' tendencies to play in science and use dispositions such as verve, rhyme, and music. Carambo highlights contradictions that occur all too often when meso level achievements do not translate into macro level transformations and accomplishments.

Sterba, in chapter 10, examines sexuality of urban females in terms of agency and oppression in a chapter in which macro, meso, and micro perspectives inform the analysis. In her use of Anderson's street code (Anderson, 1999) she explores the extent to which the necessity of urban youth to maintain respect also shapes the practices of females in the streets and in school. Through an analysis of Rashida's practices in science and mathematics, science and everyday life it is apparent that she employs sexuality to her long-term disadvantage.

In chapter 11 Scantlebury examines gender equity in the teaching and learning of science at City High. Studies of the practices of May and Ivory allow Kathryn to show how each is disadvantaged through her role as an othermother, thereby having to work hard in the home and sometimes having to miss school. In an examination of the interactions of May and Ivory in Carambo's science classes, Scantlebury shows how Ivory is advantaged by Carambo's teaching because his practices create resonances that allow her to participate fluently. In contrast, May is disadvantaged because she has different forms of capital that do not resonate with Carambo's teaching to the same extent as occurs for Ebony. Cogenerative dialogues are considered as a possible solution to equity problems.

In chapter 12 Otieno, a teacher researcher, and Milne, a university researcher, explored the uses of levels of representation as theoretical tools for improving the quality of science education at the neighborhood high school in which Tracey was teaching. The chapter highlights the interconnections between doing research and planning and enacting science curricula. The authors explore different ways of thinking about the success of activities such as the one Tracey used and questions are raised about the suitability of pencil and paper measures of science achievement.

Chapters 13 and 14 are the first of two pairs of companion chapters in which a teacher explores her teaching autobiographically and then the university researcher examines teaching and learning in her present day classes to search for the patterns identified from the historical analysis. The argument is that present day practices are historically constituted. In chapter 12 Martin explains how her life as a student was characterized by poverty and home circumstances that assured her of minority status and many occasions of oppression. She shows how she overcame many structural constraints and used capital from her cultural and social toolkit to effectively learn to teach in two markedly different urban schools. The study provides insights into the agency | structure dialectic and examines how, at both schools, Sonya was agentic in a structural arrangement that was challenging.

In chapter 14, LaVan undertook research in Martin's chemistry class at Urban Magnet, one of the highest-achieving schools in Pennsylvania. LaVan shows how Martin's agency allowed her to create structures that afford the agency of her students and allow them to work together to develop deep understandings of canonical science.

Loman adopts an autobiographical approach to her professional development as a science teacher, beginning with her experience of teaching science in Papua New Guinea as a Peace Corps participant. She then situates her teaching in New Mexico at the high school in which she was a student and then in two quite different urban schools in Philadelphia. Loman's use of activity theory allows her to examine significant issues in an historical context, chapter 15 being the second to employ biography as a research tool. These include the roles of social, cultural, and symbolic capital in teaching and learning science, the significance of connecting professional practice to the culture of the students and their parents, and the uses of material resources to build meaningful learning of students. Loman also explores gender issues in her schools and classrooms and the extent to which what is learned in science is useful in life out of school.

In chapter 16 Olitsky picks up on the themes developed by Loman and examines their instantiation in her current practices. A feature of the chapter is the use of students as researchers and cogenerative dialogues to develop collective understandings between Olitsky, Loman, and the student researchers. Olitsky uses a blend of activity theory and cultural sociology in an investigation in which patterns of coherence and contradictions are pursued in the development of social, cultural, and symbolic capital among students and between the teacher and students. Among the salient issues addressed are the role of humor in teaching and learning science, the manner in which interesting content makes a difference in the learning of science and the significance of reflection in bringing teaching and learning roles to a level of awareness such that changes can be contemplated. A concern throughout the chapter is the inclusiveness of science and its relevance to the lives of students.

In chapter 17 we identify the themes that we find most salient to resolving the problems of science education that have for so long been pervasive in science education. Implications are presented for students, teachers and administrators in schools, science teacher educators, and researchers and policy makers. In a bold look at the nature of science we look ahead and explore its potential to transform the social lives of students and in the process for science itself to be transformed.

Transcription Conventions

The research in this book is mainly critical ethnography, with methods being used to explore social life at macro and meso levels. Microanalyses also are employed, involving transcribing of audio and video tapes and adherence to a set of conventions. These conventions are featured in chapters 2, 4, 5, 6, 7, 8, 9, 14, and 16. The conventions are shown below.

(0.41)—time in seconds.

In chapter 4 a pair of square brackets, [ ], bridging consecutive lines, indicate beginning and ending of overlapping speech. In other chapters overlapping speech is shown only in terms of when it begins. A left bracket, [, is used to show where the overlap in speech commences.

RIGHT—capital letters denote louder than surrounding speech volume

°Yea°—degree signs enclose speech with lower than normal volume; *ten*— italicized utterances were stressed

↓away—down and up arrow followed by underlined words denote descending and mounting volume over the stretch of the underlined text

Gir::1—each colon indicates an extension of a phoneme by 0.1 seconds

*—the asterisk aligns speech and video offprints

=—equal sign shows latching, that is, two utterances are not separated by the normal pause

(??)—each question mark enclosed in the parentheses represents a word that could not be deciphered

.,?!—punctuation is used to indicate speech features, such as rising intonation heard as a question, or falling intonation to indicate the end of an idea unit (sentence).

Chapter 1

The Who, What, Where, and How of Our Urban Ethnographic Research

Gale Seiler and Rowhea Elmesky

Our research in schools in Philadelphia has provided us with many different experiences and with a range of images of urban schooling. We have watched sweat dripping off the faces of high school students hard at work planting trees and flowers in a schoolyard garden on a hot summer day, and we have seen students dodging raindrops as they race from the subway stairs to the school door, only to arrive late, and decide to turn back rather than receive detention. We have spent time in classrooms so hot that the teacher must use rickety fans to create a bearable learning environment, and we have attended science labs where students are bundled in winter coats and hats to combat the frigid classroom during the winter semester. We have watched students previously not considered responsible enough to use a scalpel adeptly slice connective tissue and remove a sheep's brain from its cranial case, and we have seen our own and other teachers' lessons dismantled by student behavior. We have witnessed students being expelled and taken away in handcuffs, and we have accompanied students to nationally renowned education conferences to present sessions focused on improving teaching and learning of science in urban schools and listened in awe at their eloquence.

There are many ways one could represent the inner city public schools of Philadelphia. Without being able to include a selection of video clips with this book, we instead use the above recollections to provide snapshots of practices enacted by students and teachers. These illustrate, in some cases, the richness and, in other cases, the impoverishment of material and human resources available in urban schools. While some of the images listed above are ones that would be expected, others contradict common notions of urban schools, urban teachers and students, and urban science classrooms. Typically, when conversations center on urban schooling, images of students, teachers, and the daily unfolding events are obfuscated by assessments of student performance on standardized tests and phrases such as "at grade level," "percentile," and "below basic level." Thus, while the narratives with which the chapter opens can paint an insider view of public schools in Philadelphia, more commonly, scores and statistics serve as the dominant means by which the schools and the students are represented.

What the Numbers Say and Don't Say

While numbers alone can cause us to lose much of the richness and complexity of what urban schooling is about, they certainly hold a central position in determining policies by which schools and school districts are run today. These are the measures through which schools are evaluated, and more importantly, these are the numbers through which students are judged, their achievement assessed, and their life chances, in many ways, determined. Thus, failed courses and low test scores not only provide insight into the status of teaching and learning in urban schools, they impact the students' lives by serving as "access codes" for students' success in the educational system and, ultimately, in society.

According to these quantitative measures, Philadelphia schools are generally low performing. On statewide assessments, testing reading and mathematics skills, clear differences exist between the public schools of Philadelphia and public schools throughout the rest of the state. One assessment tool, requiring students to complete both multiple-choice and open-ended reading and math questions, produces scaled scores ranging from 1,000 to 1,600. In spring of 2002, the state averages for eleventh graders were 1,320 in reading and math, while Philadelphia's averages were 1,170 in reading and 1,180 in math.

The inadequacy of Philadelphia's public schools in educating its population of predominantly African American students has been documented and intensely studied for many years, and remedies have been continuously attempted. Despite this longitudinal effort, in 2001 nearly 60 percent of Philadelphia's students still had scores that placed them in the bottom quarter of students statewide. These low scores played a part in the recent takeover of the Philadelphia Public Schools by the state under a new law.

Like other major cities (Kozol, 1991), the public schools in the Philadelphia area can be characterized as a three-tiered system composed of suburban schools, superior citywide magnet schools with high achievement (serving ethnically, racially, and economically diverse students), and neighborhood city schools with low achievement (serving mainly poor and minority students). The latter are the most common high schools in Philadelphia and are also called zoned or comprehensive high schools, indicating that each school serves students from a given set of neighborhoods that constitute the attendance zone for that particular school. These neighborhood high schools and the citywide magnet schools differ significantly from each other with profound gaps in student achievement, funding, graduation rates, and college admission and success. Recently, in response to such gaps, a new type of school (charter school) has been founded in Philadelphia, to counter the failure of the low-achieving public schools.

In this volume, we write about four schools located in Philadelphia: Urban Magnet (a citywide, selective admission middle and high school), City High School (a comprehensive neighborhood high school in the western part of the city), Southeast High School (a comprehensive neighborhood high school in the southeastern part of the city), and Charter High School (a new charter school). Great differences exist within the Philadelphia school district, particularly between the selective admission schools and the neighborhood schools. African American students are underrepresented at the five selective admission high schools, comprising 47 percent of the students at these schools but 65 percent of the students in the city's public schools. Urban Magnet, the premiere school in the city, serves a student population with 39 percent from low income, while at City High and Southeast High, both comprehensive high schools, the poverty rate is 80 percent and 71 percent respectively. The percentage of students graduating in four years from the comprehensive high schools in Philadelphia ranges from 26 percent to 66 percent while Urban Magnet boasts a graduation rate of 100 percent. Graduation rates for the new Charter High are not yet available.

Philadelphia is a city with great residential segregation in its neighborhoods. Thus the schools that serve these neighborhoods are strongly segregated as well. This is evident in the two neighborhood high schools that we write about in this book, City High and Southeast High. According to the 1990 census, West Philadelphia as a whole is 72 percent black and 98 percent of the students at City High are black. Southeast Philadelphia is a patchwork of smaller neighborhoods of varying races and ethnicities and this is represented at Southeast High where

42 percent of the students are African American, 27 percent are Asian or Asian American, 24 percent are Caucasian, and 7 percent are Latino. For almost half a century a plethora of studies, reports, reforms, and special projects have attempted to ameliorate what the Report of the Special Committee on Nondiscrimination (1964) described as the "educational lag in the predominantly Negro neighborhoods" (29). The report noted the impact of de facto segregation and its incumbent inequities on the quality of educational opportunities and achievement of African American students. One major structural change made to Philadelphia schools was introduced in 1988 in the form of a funded initiative to create smaller, self-contained learning communities within schools (Fine, 1994). These schools-within-schools were initially known as charters and later as Small Learning Communities (SLCs). Though the establishment of SLCs improved some aspects of the schools, other problems were not remedied and unexpected issues arose from this new structure. Many of these developments are illustrated in chapters 2 and 3.

Despite reports and reform efforts, issues of poor achievement persist in Philadelphia, decades later. A headline in the summer 2000 edition of the *Public School Notebook* (an independent quarterly newspaper) read "Neighborhood High Schools: Results Are Still Alarming." For instance, in 1988, less than 50 percent of ninth grade students in Philadelphia's comprehensive neighborhood high schools progressed to tenth grade (Fine, 1994), and only 36 percent of the 1993 ninth graders at comprehensive high schools in Philadelphia graduated in 1997. Graduation and enrollment figures for the neighborhood high schools, today, show that most students who fail to graduate usually do not get beyond the ninth grade. At Southeast High, 44 percent of the students are freshmen and only 13 percent are seniors. Moreover, it is estimated that only one out of every five students who attend a neighborhood high school graduates with the course credits needed for college and many of these are poorly prepared for college in substantive ways (Whitman, 2000).

City High

City High is located in West Philadelphia, a largely black neighborhood just a few blocks away from a major university. A trademark mural marks the entry of the school, telling the story of the destruction of the houses and neighborhood that once stood on the school site and depicting the controversy surrounding City High at its inception. When opened in the early 1970s, City High was caught in a three-way tug of purpose, between the city's need to reduce overcrowding in schools in West Philadelphia, the desire of university and business interests to develop a science Urban Magnet school, and the community's interest in improved educational opportunities for the local youth. What emerged was a compromise, which prevented the full attainment of any one of the original three goals (chapter 3).

City High has struggled through three decades, and the complexities and contradictions of its origins are still currently visible in scars, which are detailed in chapter 3. The number of SLCs has fluctuated between a high of ten to six at the current time, as reorganization has taken place in response to new problems and school system directives. The SLCs at City High have traditionally had themes to address student interests such as the health professions or communication. City High serves a student body of over two thousand students, most of whom are black and many who are from conditions of poverty, and these students continue to perform poorly on state-mandated achievement tests.

Southeast High

Southeast High was constructed in the early 1900s and served as a middle school until the late 1980s. Thus, it lacks many of the resources found in the surrounding suburban high schools or in newer high schools in Philadelphia. In looking at the school facilities, there are many reminders of its recent history as a middle school. For example, the cafeteria tables and chairs are close to the ground, intended for much smaller people. The library and gym are also roughly half the size of their counterparts in high school buildings. The physical structure of the building shows many signs of its age and the lack of material resources—for example, when there is inclement weather, halls and classrooms are cluttered with buckets and trash cans to catch dripping water from leaky roofs.

As a neighborhood, comprehensive high school, Southeast High serves mainly students from the surrounding communities, although it also accepts students who apply from all over the city. The school of one thousand two hundred students is divided into three small learning communities and an academy for law. Unlike the SLC structure in other schools such as City High, there are no specific academic requirements, ability tracking, or career-related themes (except for the law academy) for the SLCs. Standardized test data for the science Stanford Achievement Test (SAT-9) taken in 2000 show that only 0.2 percent of the school population scored "proficient," 9 percent scored "basic," 69 percent scored "below basic," and none were "advanced." Poor attendance is a serious problem at Southeast High, and 22 percent of the students were not tested due to absenteeism.

Urban Magnet

Out of the five city-wide, selective-admission public high schools in Philadelphia, Urban Magnet is the most esteemed. Urban Magnet was established in 1958, and then was subsequently expanded and reorganized. It now includes a middle school (grades five to eight) and a high school (grades nine to twelve). One thousand one hundred and fifty students attend Urban Magnet and they are admitted from throughout the city based on academic performance, standardized test scores, teacher recommendations, and attendance. Students are admitted at fifth grade, but then must reapply to enter the high school, and admission to the high school is more selective than at the fifth grade level. Only one hundred ninth graders (approximately 50 percent of the eighth graders) are accepted. Urban Magnet is the premiere public school in the state as well as the city and is nearly always at the top of the list on statewide achievement tests. According to the school website, all graduating seniors attend college, usually four-year institutions. As listed on the School District of Philadelphia's website, Urban Magnet is comprised of about 34 percent African American students. Thus African American students are underrepresented at Urban Magnet, relative to the district as a whole, as are students from conditions of poverty. However it appears that even these statistics overestimate the diversity of the high school portion of the school, which is considerably less poor and less black. Separate statistics for grades nine to twelve at Urban Magnet are not readily available, but one teacher at the school points out that there are only six African American male students out of approximately one hundred in the tenth grade.

Charter High

In 1997, the Commonwealth of Pennsylvania Legislature approved the existence of charter schools, public schools that operate independently from the local district administration. In Philadelphia, as in other locations, the charter school movement has grown out of dissatisfaction with the lack of success observed in local public schools. According to the mission statement on the school website, Charter High was "created to address the failure of the traditional high schools to prepare urban students with the academic and critical thinking skills required in 21st century society." Charter High opened in September of 2001 with one hundred students and moved to a newly renovated location in Fall 2002 with one hundred additional students. The school was created by a coalition concerned with preparing students with the skills they need to succeed in today's economy. In an effort to allow no student to fall between the cracks, students progress to the next course only after mastering the previous course material. The sequence of courses is built on a vision for developing academic literacy skills and process skills along with conceptual understandings in all content areas. These are mapped across all courses and disciplines and teachers are expected to adhere to them. Classes are generally small and personal attention is high. Thus, human resources in the form of contact with teachers and staff is greater than at traditional schools. Frequent counseling and monitoring of student progress by an individual or team of staff members is the norm. Admission to Charter High is non-competitive and not contingent on students' grades or recommendations, though it is likely that there is some self selection since the students and their families have gone through the process of applying. In terms of their academic achievement during middle school, Charter High students' performance is similar to that of the student population at a neighborhood comprehensive high school such as City High. However since the school attracts students from throughout the city, they are somewhat more racially and ethnically diverse than the neighborhood schools. For example in a biology class of twenty students, fourteen were African American, four were Latino, one was Asian American, and one was Caucasian.

While receiving the standard per capita funding from the school district, Charter High has also carried out an extensive capital campaign and has raised over $5 million from individual donors, businesses, and foundations, thus material resources are more prevalent than at the neighborhood high schools. Charter High's administration is acutely aware that in order to maintain its charter from the Commonwealth and to continue to garner external funding, it is imperative that they show improved student learning, and this is judged largely by performance on state-mandated and other standardized tests.

Methodology

In 1997, Kenneth Tobin together with a small number of coresearchers began doing research in science classrooms at City High, a predominantly African American school serving students from conditions of poverty. This inner city, poor-achieving high school soon became the center of the university's science teacher education program as well as the central research location for our studies of urban science teaching and learning. This book is based on the research that began six years ago at City High with just a handful of individuals and continues today with a group of about twenty coresearchers across four schools in Philadelphia, as well as in other cities. Our research represents not just an attempt to understand the teaching and learning of science in urban schools that serve economically and socially marginalized minority students, but to change it. Thus, the chapters in this book represent our evolving understandings of what occurs in urban science classes that leads to the low achievement of inner city students and what can be done differently so that science teaching and learning has the potential to be transformative to the lives of the students.

During the first years of our research at City High a number of questions emerged:

1. How can we do a better job of teaching science to inner city African American students?
2. How can we teach these students in ways that are transformative?
3. How can we better prepare teachers to teach students like these?
4. How can our research be catalytic in improving teaching and learning in inner city schools?

To answer these questions and accomplish these goals, we employ new approaches in our research that do not rely on traditional roles nor attempt to achieve objectivity. Our ontological views are such that we recognize that our representations of what happens in urban science classrooms are neither truth nor reality. Rather, they are our individual and collective constructions and understandings of our experiences and observations. As such, they are changeable and vary between participants. They are shaped by our schema, values, and expectations and are strongly linked, not only to the social and cultural contexts from which they arise, but also to the social and cultural resources that each participant brings to the classroom. We acknowledge this inescapable cultural subjectivity and perspective, and we make it a core goal of the research to examine and deconstruct the cultural lenses with which we, as teachers, researchers, administrators, and students, interpret classroom (and other) events and interactions. As the research has grown to include more participants and more schools, our approach has been molded by the stakeholders at each school site and in each classroom, yet there are some common threads that run throughout, as will be made explicit in the remainder of this chapter.

The Nature of the Research

Critical Ethnography

> The very nature of the educational practice—its necessary directive nature, the objectives, the dreams that follow in the practice—do not allow education to be neutral as it is always political. The question before us is to know what type of politics it is, in favor of whom and what, and against what and for whom it is realized. (Freire, 1993, 22)

As argued by Roman and Apple (1990), we live in an unequal and unjust society and our relations are deeply patterned by forces of oppression. Schools are not only shaped to a great extent by the dominant political, economic, and cultural forces, but they also play a powerful role in the perpetuation of an unjust system. Thus, schools, their curricula and pedagogy can not be considered neutral, particularly for African American students and other disadvantaged groups in the nation's largest urban centers. In fact, decades after the desegregation of American schools and appreciable gains in civil rights, the condition of science teaching and learning within inner city schools serves as evidence not only of this country's grim racial past, but of persistent modern day social inequities and the continual struggle of African Americans and others to vacate their disadvantaged position in society. While many minority students possess real hopes that their lives can improve through schooling, in reality, there exists a cycle in which the current socioeconomic position is reinforced by and simultaneously reinforces their experiences in the educational system. The transformative potential of schools and education has not been met and instead oppressive structures are reproduced.

Our research in urban schools in Philadelphia is concerned with the reproduction of social classes and racial oppression and, in particular, with the role of urban schools in this reproduction. As we believe that there is no neutral education, we also argue that there is no neutral research in education (Lather, 1986). In considering education as well as educational inquiry as political and ethical acts, we take the position that educational researchers are obligated to enact research methods that catalyze social transformation. With this goal, we adopted a methodology through which we could change the structure of science education research, particularly the ways in which it was traditionally enacted when studying urban schools (Barton, 2001). We embrace a critical ethnographic research methodology that affords the agency of those involved such that the research process and results provide a catalyst for the participants' growth and ultimately for structural transformation. Whereas noncritical research in many ways perpetuates the status quo, critical research is concerned with unmasking dominant social constructions and their interests, studying society with the goal of transforming it, and freeing individuals from sources of domination and oppression. Moreover, critical ethnographic research methodology, which draws upon feminist research, neo-Marxism, and Freire's "emancipatory" theory, requires educators to make a political commitment to the struggle for liberation of the oppressed, and recognizes the embeddedness of knowledge and culture in economic and political interests. In discussing the role of critical research in science education, Kincheloe (1998) draws from the Frankfurt School of critical theory. Though there are numerous varieties of critical postmodern social theory, he describes critical research quite simply.

> The "critical" aspect of critical research assumes that the inequalities of contemporary society need to be addressed and that the world would be a better place if such unjust realities could be changed. Thus we explore the world, science included, for the purpose of exposing this injustice, developing practical ways to change it, and identifying sites and strategies by which transformation can be accomplished. (1191)

Thus, our critical research goes beyond interpretive and naturalist research in our uncamouflaged attempts to enable students to alter their social positions and work against social reproduction and for teachers to broaden their cultural perspectives.

Catalytic. Critical ethnography not only provides a methodological framework to document and analyze the discriminatory practices supported by schooling (particularly urban schooling), it also necessitates that the research approach must be catalytic in altering those practices. We are particularly concerned with the catalytic nature of our research since we work with populations who are traditionally silenced. For decades, others have spoken to the needs of both teachers and students and decided what should be done within low-performing schools. Urban youth, teachers, and administrators in those schools may want to instigate change but lack the power to do so in any meaningful ways, since reform initiatives, as well as the research processes driving the initiatives, bypass those people who are most directly impacted. Since individuals' social constructions function to sustain a hegemonic system of social oppression, our research approach intimately involves other stakeholders as we try to unveil those constructs and critique them. As described in later chapters, agency, or the power to act, is dialectically related to structure (Sewell, 1992). Thus, we adopted a research approach that affords stakeholders the opportunity to further their self-understandings or "conscientization" (Freire, 1970). In this way, the process can be catalytic for those involved and transformative to structure. The researchers who present their work in this book show a variety of ways in which the research process has been catalytic as they address their awareness of oppression related to the nature of schools, pedagogy and curriculum, the nature of traditional research roles, and the nature of science.

Changing roles in research. As noted by the National Research Council (1996), reform initiatives have swept through our schools, year after year, as we strive as a nation to meet goals of academic excellence. However, the quick coming and going of educational reforms often elicits criticism from practitioners in schools, even more so when they are based upon "research-on" rather than "research with" methods. While some practitioners may be interested in trying suggested or imposed teaching approaches, more often they feel alienated from the decision process that determines what constitutes good teaching and learning (Cochran-Smith and Lytle, 1990). The potential for our research to move beyond this and to be catalytic has been increased by certain attributes of our research group's structure and dynamics. In order to optimize potential for change, we believe that the roles of the participants must extend outside the traditional labels of researcher and researched. Thus, the coresearcher hat is worn by many participants ranging from students, classroom teachers, school administrators and student teachers to university master's students, doctoral students, postdoctoral students, and university professors. Cogenerative dialogue forms a core component of our efforts to share power among students, teachers, and university personnel. Moreover, qualitative data are collected, analyzed, and understood by this variety of stakeholders, and all participants are contributors, learners, and researchers and are afforded the opportunity to have their goals met, though there may be different sets of goals. Hence, in this book, the writings represent a great diversity of voices and perspectives among participants in the research, including practitioners, school personnel, university personnel, and students.

Students as researchers. We have found the involvement of students as researchers to be a crucial component of conducting ethnographic research that looks at the improvement of science teaching and learning (e.g., Tobin, Seiler, and Walls, 1999; Elmesky and Tobin, 2003). Over the span of this research, we have involved student researchers in many different capacities that engage them in tasks far beyond member checking. In fact, the influence of students in our research is so pervasive that their mention in this chapter substantially understates their many contributions. They help produce a variety of artifacts for our analyses, are an integral part of many of the interpretive processes, and act as teacher educators, curriculum developers, and ethnographers. Some of their contributions include: journal entries, transcriptions, self-authored raps, data from internet research, video analyses, ethnographies, independently designed interviews, the production of science-related videos and presentation materials associated with seminars and professional conferences. They regularly participate in cogenerative dialogues (Roth and Tobin, 2002) in which they are introduced to sociocultural theoretical lenses and practice applying those constructs to make sense of data resources. Student researchers have also coauthored and copresented papers with us at professional conferences. Although no chapters in this volume were authored by student researchers, they have written the book foreword, and their voices and perspectives are represented by the authors in other ways, as well. For example, with the assistance of student researchers, Elmesky, Seiler, and Sterba (in chapters 5, 6, and 10) have been able to more authentically address science education within a larger context of the home and neighborhood.

A key resource to our understanding of the ways in which we can improve teaching and learning in urban science classrooms has been student researcher ethnographies of social spaces in which they participate outside of school. Once constructed by the students, they become a resource for further joint analysis by researchers and students. Thus, student researchers provide us with access to additional data resources from which to identify students' patterned practices and the associated contradictions in fields not typically accessible to teachers—neighborhood streets, homes, work places, and basketball courts, for example. In this way, a realm of urban education, virtually unexplored in educational studies, has been illuminated. For instance, during the summer of 2003, four student researchers created personally edited video ethnographies of their lives. They videotaped a variety of social fields specific to each of them and then utilized sociocultural theoretical constructs along with technological resources (video editing equipment and software) to identify video vignettes of their choice. Searching for patterns of coherence and contradiction, the youth focused on producing an edited version of "who they are," including the embodied dispositions that make them "who they are."

Recursive: Research-theory-practice. In addition to student researchers, our research group is particularly committed to the involvement of teacher researchers in a research model that links research, theory and practice. In our work, university-based researchers form teams with teacher researchers and student researchers at each school site to work together in trying to understand the classroom events including teacher and student actions. Moreover, the university-based researcher does not stand on the side to watch and judge the teacher researcher's classroom, rather he or she coteaches with the teacher researcher in order to gain an emic perspective of "we" rather than "us versus them." Cogenerative dialogue serves as a forum for linking research, theory and practice (Roth and Tobin, 2002) by introducing sociocultural theory and applying theoretical constructs to identify and examine poignant video footage and transcripts of interactions in classrooms and other social spaces. These focused discussions between classroom coparticipants promote a strong sense of shared responsibility among the stakeholders and allow both overt and below the surface actions to be examined. Fueled by the introduction of sociocultural theory as a resource for understanding classroom practices, cogenerative dialogue allows classroom events to be a source for theory-building as well as for inviting changes in classroom practice. Most importantly, cogenerative dialogue provides a structure for maintaining mutual focus or entrainment and fostering solidarity and positive emotional energy (Collins, 1993) as we form collective understandings among researchers (chapter 8).

In this book, several chapters illustrate how teachers' involvement in research has impacted their classrooms and the teaching and learning of science. For example, chapters 14 and 15 are each individually authored by a teacher researcher (Linda Loman) and a university-based researcher (Stacy Olitsky), respectively, and together illustrate the interconnected nature of research, theory, and practice in the work that we do. Not only have many of our research findings directly contributed to change in the classrooms of those teachers involved in research, the findings have been disseminated within the teacher education programs at the universities with which we are associated.

Emergent research. Having a model that encourages connections between practice, research, and theory necessitates a research design that is emergent in nature. Whereas, we began with general interests in improving the teaching and learning of science in urban schools, we did not have a master set of procedures and protocols to which we adhered. The questions listed earlier in the chapter, as well as others, emerged as we became immersed in the schools and in the research. Thus while the questions initially came from the university personnel, that quickly changed as more stakeholders began to contribute to the processes of data collection, data analysis, theory building, and changes in practice. For example, early on in the research, the importance of respect repeatedly arose from student researchers in interviews and cogenerative dialogues, and it soon became a key construct for all of us. In order to be better able to recognize which teaching practices could be perceived as respectful or disrespectful by students we needed to understand the schemas associated with respect in fields outside of school, and this required the evolution of new approaches, new uses of technology, and new roles for student researchers. With these tools we began to see how conflicting understandings of respect impacted the classroom structure in ways

that could either afford or truncate individual and collective agency. We also became increasingly focused on the structure of fields outside of the classroom to
better understand why particular practices associated with respect or disrespect
were appearing in the classroom.

Quality of the research. The trustworthiness and authenticity criteria, provided by Guba and Lincoln (1989) for judging the quality of inquiry that is not
of the positivist paradigm, have been important constructs for guiding and monitoring our research methodology as it has evolved.

The trustworthiness of our research can be considered through the criteria of
credibility, dependability, and transferability and these are embedded in the nature of our research approach. Several techniques are employed for increasing
credibility. First, analytical depth and scope are made possible by the persistent
and prolonged nature of the research process. Owing to longitudinal experiences
with our researchers and within research sites, we are able to "overcome the effects of misinformation, distortion, or presented 'fronts,' to establish the rapport
and build the trust necessary to uncover constructions, and to facilitate immers-
ing oneself in and understanding the context's culture" (Guba and Lincoln, 1989,
237). In addition, our practice of engaging in cogenerative dialogue helps to establish the credibility of assertions since they are co-constructed by multiple
stakeholders as alternative interpretations are explored. This is particularly important since the urban student researchers are largely from very different lifeworlds and perspectives than the other research participants. Our approach to
comparative interpretations differs from traditional member checking in which
interpretations are simply confirmed or denied by those involved. Furthermore,
practices such as cogenerative dialogue and team research meetings allow us to
monitor our evolving constructions, in a way referred to as progressive subjectivity. Peer debriefing is accomplished both within and outside of the research
group. By sharing and examining research findings within a biweekly seminar,
during prolonged writing tasks, and at national conferences, we consistently engage in peer debriefing to express the changing constructs and to receive feedback that further sophisticates our understanding. Moreover, since our theoretical framework views culture as weakly bounded, we systematically search for
patterns of coherence and contradiction both within and across fields, not just as
discrepant cases but as opportunities for change. In fact, it is through the docu-
mentation of our methodological and analytical shifts and changes that dependability in our research is accomplished. Finally, transferability refers to the
"reach" of the research and to the possibility of the research findings being useful to those who will read them or encounter them in other forms. The temporal
depth and breadth of the research, and the use of audio and video recording, in-

fluence the detail and richness with which we can communicate our findings and the context in which they were formed, thus aiding the appropriateness of transfer. That we involve stakeholders from various positions within the education system also permits the findings to be potentially significant to a variety of individuals.

The criteria for ensuring the authenticity of the study are ontological, educative, catalytic, and tactical authenticity in addition to fairness (Guba and Lincoln, 1989). The very structure of our inquiry supports authentic research since we solicit and subsequently honor stakeholder constructions through cogenerative dialogues and related sessions. These types of activities provide opportunities for the negotiation and co-construction of recommendations and subsequent actions, and hence contribute to fairness in the research. The collaborative types of structures and practices, that form the heart of our research, allow for the creation of individual constructions that become more and more sophisticated (ontological authenticity) and afford opportunities for understanding the constructions of others (educative authenticity). In actuality, this very process and its outcomes are studied as we examine the development of emotional energy (Collins, 1993) and collective responsibility.

Since our research is concerned with social transformation and agency, mechanisms for action and change are deeply embedded in all aspects of data collection, analysis, and representation of findings, as we monitor the extent to which the research is transformative to science teaching and learning (catalytic authenticity). The significance of the catalytic effect is even greater if it empowers stakeholders to act to make changes in structure (tactical authenticity). These last criteria are arguably the most important and demanding measures of the goodness of qualitative research, and the subsequent chapters will demonstrate the extent to which individual, collective, and institutional transformation has been catalyzed and realized.

Analysis of Data

Our approach to data analysis and interpretation differs from traditional ethnography in two major ways. First, since our theoretical lenses call for the recognition of patterns of both cultural coherence and contradictions (Sewell, 1992), our approach values both of these, and whenever we make claims, we search for and expect to find contradictions to our claims. Second, we employ a multitiered approach for making claims, and for supporting and contradicting those claims, which span micro, meso, and macro levels.

Since we are interested in countering social reproduction and studying agentic practices that can be transformative for students, our ultimate concern lies with transformation of macro level phenomena. However, as the chapters in this book illustrate, it is impossible to understand interactions and events at a macro level without performing analyses and making associated claims that extend across the other two tiers (meso and micro). Hence, we see our ethnographic research as necessarily incorporating a methodology that embraces analyses and interpretations that are multileveled. The following section provides insight into how we employ different lenses for zooming in and out (Roth, 2001) and moving across levels to arrive at macro, meso, and micro level claims and understandings.

Micro, Meso, Macro Levels

Micro, meso, and macro lenses may overlap or contradict, yet employed together they render an understanding of the complex and multifaceted nature of the social realities in urban classrooms. When we make claims at one level in our research, these are supported and contradicted with analyses from other levels as well. Meso level claims rely upon traditional ethnographic analysis, and emerge when everyday unfolding events are captured as data resources, through recordings, field notes, journaling, or interviews. They arise from being in the classroom or from watching videotape footage at real time speed. In our research, meso level observations help us to identify major themes to explore for further understanding. For instance, in noticing students' tendencies to move rhythmically, and to tap out musical beats, sing or rap during science class, we examine these instances closely and look for similar instances that are played out to varying degrees. We attempt to identify instances when these practices appear, with whom they occur, how the teacher and other students react, and how they contribute to or hinder the development of dispositions associated with the science culture. Patterns and contradictions identified at the meso level, are often further explored through micro level analysis in which we attempt to understand the more detailed aspects of interactions within a particular field.

While we see unfolding meso level data resources as vital to acquiring a "feel" of what is occurring in urban science classrooms, our research methods draw extensively on microanalysis to make sense of the details of what is happening, where classroom practices come from, and what they contribute. At this tier, transcriptions are enhanced by the notation of pauses, overlapping speech, emphasis, and volume. Microanalysis also entails studying vignettes by viewing videotapes across a range of speeds, from slow motion to frame by frame. This type of analysis often reveals fleeting actions, subtle movements, peripheral events, and nonverbal communication that are not easily identified in real-time viewing. Specifically we may look at facial expressions, direction of gaze, hand movements, body position, and use of material resources. In this way micro-analysis is a methodological tool that is consistent with our theoretical valuing of unconscious aspects of practices. These micro techniques expand our capacity to explore minute aspects of student and teacher actions and interactions that support, as well as contradict, our meso and macro level observations.

In another example from our research, we became interested in understanding the role that social capital plays in classroom interactions and its relationship to students' building of science cultural capital. In order to develop understandings that are informed by microanalysis, we look for vignettes where students exchange social capital and instances when the building of social capital appears to be linked to the building of science cultural capital and when it is not. Once such meso level vignettes are selected, we then rely on microanalysis to help us identify gestures and verbal and nonverbal cues that are utilized as students exchange social capital with each other, and how these subtleties aid or hinder science learning.

We define macro claims as ones that extend across social space, time, or both space and time. One way we enact macroanalysis is to increase the speed of video viewing up to five times normal speed. This allows us to advance time as we examine patterns in temporal changes in actions, interactions, and the use of space and material resources. Our concern with social reproduction invites us to look at students over time (longitudinal engagement) and across social fields (including but not limited to the classroom) for evidence of whether science is a transformative force in the students' lifeworlds. For example, at times we focus on macro level analysis across social spaces by searching for student practices seen in urban science classrooms, but also seen within other fields such as the neighborhoods, streets, and work places (Elmesky, 2003). In addition, many of our research claims are based on analyses that extend across several semesters and, often, span years.

It is important to state that our ability to make macro level claims and find supporting or contradicting evidence regarding science teaching and learning is deeply connected to the inclusion of student and teacher researchers in our research teams as well as our longitudinal involvement with them.

Without the participation of student researchers we would not be able to access their experiences school-wide, beyond the classroom in which research is being conducted, nor would we have insight into their practices in different localities outside of school. Thus, we have involved student researchers across a number of years, have had the opportunity to interact with them in a variety of social fields, and rely on them to provide data from their lifeworlds. Over time we are able to keep track of macro level changes in their lives such as moves to new cities, different schools, graduation, college plans, success in school, failed grades, and employment opportunities. Moreover, they help us understand how their involvement in the research has afforded agency in fields temporally and physically separate from school.

The longitudinal commitment by the teacher researchers is also an important part of the study. Many of the authors included in this book have been involved in the research across important parts of their life courses as teachers and learners. For example, Jen Beers, the author of chapter 7, began to work with us when she was a student teacher at City High. Now a second year teacher at Charter High, she continues to carry out research in her classroom and to involve new student participants in the work. This type of longevity enables us to also understand some of the impacts of involvement in this type of research on the teaching and life trajectories of teachers as their schema and practices change.

Looking Forward: The Chapters to Come

The range of authorship and styles in the following chapters illustrates the varying and emerging roles that the research participants have had and provides multiple perspectives across levels within the schools and university. Thirteen chapters represent studies of inner city science classrooms based on data collected through collaborative research. Eight of the thirteen chapters are located in City High, four are associated with Urban Magnet, and one is situated in the context of Southeast High. As mentioned previously, one chapter chronicles the author's journey from student teacher at City High to classroom teacher at a new, urban charter high school. Three chapters grew out of science teacher autobiographies. A number of the chapters are written wholly by classroom teachers while two are written collaboratively by a classroom teacher and a university researcher. In some cases research was carried out jointly by a teacher researcher and a university coresearcher but the collaborators decided to represent their work independently in writing separate companion chapters.

Editors' Perspectives

Improving Urban Science Education embraces a methodology we believe to be central for research in science education; one that is catalytic and that furthers the interests of participants, especially those who may be bypassed by the opportunities education purports to offer all learners. The title of our book is both optimistic and a challenge, embracing the agency of urban youth who, through their studies of science education, can produce and reproduce forms of culture to allow them to take full advantage of a world that is waitin'. However the title also challenges the world to be ready for them, and open to them. Our research, in which urban youth and their teachers are active coparticipants, seeks to transform urban science education. Through our research, we expect to identify forms of practice whereby students and teachers collaborate to transform science education in inner city schools. Our conviction that student and teacher practices are central in reforming science education departs from traditional approaches in which well-intentioned outsiders mandate what should be taught and learned and hold schools and teachers accountable for the achievement of learners. A world waitin' implies that what urban youth learn from science education creates opportunities for them to improve the quality of their lives in and out of school. Accordingly, our research includes macro questions on how and the extent to which urban youth use what they can do and know of science to create productive learning environments at school and bold vistas for their lives out of school.

Chapter 2

Urban Science as a Culturally and Socially Adaptive Practice

Kenneth Tobin

Control the Students

In the early days of television in Australia, Vin Walsh was the weatherman. Also, he was a teacher educator. As a seventeen-year-old in the 1960s, I gathered with my family on most Friday evenings to watch live weather reports as Vin explained the forecast for the weekend in terms of the highs, lows, advancing fronts, and the behavior of his bees. Imagine my excitement when I went to teachers' college, to have the "bee man" as the teacher of a class on general teaching methods. This class was exactly what I wanted. An expert on teaching (and weather and bees) offering wisdom on how to teach. "Find the toughest kid and then ask him to pick up a piece of garbage from the floor (show them who's in charge). Don't speak until they quiet down. Wait. They will then become silent (control pause). If kids are misbehaving walk near to them, but don't stop teaching (proximity desist)." I regarded Vin Walsh as authentic and, during field experiences, I wanted to implement his suggestions since he would observe and evaluate my teaching performance. Surprisingly, most of what he recommended worked just as he said it would and his tips became part of my repertoire of teaching strategies. Embedded within these suggested teaching strategies, and most others he volunteered, was an implicit requirement that the teacher establish control over students as a prerequisite to being able to teach them. The creation of productive learning environments was premised on teachers being able to control students and thereby require particular types of participation.

In almost a decade of high school teaching in Australia I relied on my youth, high levels of energy, and growing expertise in science to create learning environments that were effective for teaching predominantly white, middle-class, Australian youth. In a nutshell, in the early weeks of a school year, I established control, adopted a clear rule structure, and taught in a no-nonsense style that gradually provided more autonomy to students. I was able to command respect from my students and, by being fair and consistent, I experienced few problems. Accordingly, I expected to be successful when I decided to teach science in a large comprehensive neighborhood high school in Philadelphia as part of my research on urban science education. However, there were reasons for caution. Socially and culturally my students were markedly different from me and the students I had taught in the 1960s and early 1970s.

Urban High Schools—Northeast Style

Even though it was a brisk fall morning in 1997, the images of my initial visit to City High are as vivid as if it occurred yesterday. At that time, the route from my office to City High was unfamiliar and I was reliant on a colleague to get me to the school. As we navigated narrow streets to approach the school from the west, my mind was replete with deficits. The streets were littered and as we approached the school, a rusted chain link fence enclosed what appeared to be a prison compound, with a vacant expansive field separating the fence from a cluster of dour, brick, multistory buildings. "K to eight elementary school," my colleague explained as we made our way along an uneven pavement toward a much larger building on the corner of a bustling city street. "The school has real problems. They've had several principals in the past year. No stability," he continued. I nodded grimly, not sure how to respond. I had just moved from Florida to Philadelphia to undertake research on urban science education and this was my introduction to urban schools, northeast style.

The large building was City High. Inwardly I groaned as I anticipated what lay ahead. The three-story structure had few windows facing the street and a metal grid covered those that were there. I peered through the grid that protected ground level windows and could see the dim form of a basement level classroom. As we headed for the steps, I mustered up images of what was inside, and felt anxious as we made our way toward the entrance, behind six heavy metal doors. I could not imagine being a student or teacher in a school like this. By trial and error we found a door that could be opened and approached several security personnel (called nonteaching assistants or NTAs) who checked our ID cards, instructed us to sign in, and directed us toward the office.

Although the school year had not yet started, the office was chaotic as parents and students waited their turn to speak to school officials. My colleague knew his way around and we didn't have to wait. "Where's Rusty?" he inquired. An office assistant nodded a greeting and motioned to the door with a sweep of her hand, "In the operations room."

The operations room was in the basement, formerly a metal trades workshop, now used as a "bunker," a place for the principal and other administrators to meet, plan, and escape the bustle and inevitable interruptions of the main office. I was impressed with Rusty as he greeted us and then took us on a tour of the school. He demonstrated an inspiring capacity to code switch as he addressed nonteaching assistants at the front door, parents, students, office assistants, and us. A graduate with a doctorate from an Ivy League university, he was self-assured and knowledgeable of research, theory, and the latest developments in school organization and leadership. Rusty had recently transferred to the school from the school district office, a demotion he attributed to philosophical differences with a new superintendent. Rusty relished the challenge of being principal of a large inner-city high school consisting of over two thousand students, most of whom were African American and from home circumstances that were economically challenging.

As we walked, Rusty described the history of the school. In the late 1960s City High was planned as a state-of-the-art magnet school for science. However, due to community opposition and budget constraints City High opened in 1972 as a comprehensive high school. Almost immediately, rivalry between gangs and racial conflict created safety problems, which continued for more than two decades. When Rusty became principal his highest priority was to reorganize the school to be safe. Accordingly, subject area departments, which Rusty considered unworkable, were discontinued and a trend to create Small Learning Communities (SLCs) in the school was strengthened. The chief idea was that students would experience a small school feeling, go to classes in a confined part of the building, and be taught by a relatively small number of teachers who could personalize the experience. Each SLC had its own space in the building, where students took all their classes. A downside of this policy is that, while science was taught throughout the building, the specialist labs and resources needed to support science education were only found on the third floor. Spread throughout that level were large laboratory spaces, preparation rooms, storage facilities for science supplies and equipment, and many science classrooms that included combined laboratory and classroom facilities. Almost all of the rooms had the option of being able to slide back a common wall, with an adjacent room, to create a larger space for teaching and learning.

Surprisingly, at the time of our tour, a relatively small fraction of the rooms on the third floor was in use, mostly for courses other than science. Only four rooms were used for science teaching, and one was set up so that the lab assistant could schedule students and teachers from other floors to participate in laboratory activities. Most of the science rooms were vacant and the power, water and gas were disconnected.

The tour with Rusty raised doubts in my mind regarding the purposes and forms of science education in urban schools. In particular, as we explored the physical and social spaces of City High, I was struck by my otherness. The demographics of the school, where almost all of the students are African American, were strikingly different from those of the school district where the racial distribution of the two hundred thousand students is approximately 66 percent African American, 14 percent Hispanic, 15 percent Caucasian, and 5 percent Asian American. In addition, only about one third of the teachers at City High are African American (about the same as the percentage in the district). My concern with such demographic data was that since the social and cultural backgrounds of the majority of the teachers were so different from those of their students, they might not know how to connect their teaching to the cultural capital of the students. Following my visit to the school, I had strong concerns about my own social and cultural history and the challenge of connecting with the youth I planned to teach. What troubled me was how to identify what students brought with them to school that could be used to build knowledge of science and what I could do to mediate the students' learning of science. I felt a need for new theoretical lenses to examine science education, urban youth, and the ways in which teaching and learning might lead to the emergence of science as a socially transformative discourse.

New Windows into Science Education

Although the search for fresh theoretical windows was not a new activity for me, following my visit to City High, I was more conscious of the need to expand on the psychological models that had traditionally served science educators. I began by exploring the capacity of social and cultural theory to raise issues and generate solutions with their associated implications for the practice of science education in urban high schools. Over the years, three frameworks, cultural sociology, activity theory, and the sociology of emotions, have emerged as valuable to my developing understandings and have been applied in a program of research that is presented in the remainder of this chapter and in the other chapters that comprise this book.

Learning as Cultural Production

The research in this book examines social life at the micro, meso, and macro levels. In the first chapter these constructs were introduced in a methodological sense, and here I address them theoretically. The micro level involves the interactions of individuals with resources and the unfolding of action, as praxis. Agency is enacted at the micro level as structures are accessed, appropriated, and reshaped. The meso level occurs within a single field and involves social life as it is experienced by participants. The macro level involves participation in more than one field, including schools, where fields are nested within one another and intersect to create complex organizations. Accordingly, students and teachers participate simultaneously in more than one field and experience agency | structure relationships that are reflective of those fields (chapter 4). The extent to which capital produced in a field can be used to reach goals in other fields is a critical issue in our research. Whether or not fields intersect, students, teachers, and other stakeholders cross boundaries, and what is learned and done in one field can afford and constrain social life in other fields. In this book, studies focus on the enactment of culture from other fields, the extent of its success, and the contradictions experienced by participants because fields have porous boundaries, thereby allowing "outside" culture to be enacted in a field, sometimes with unanticipated consequences.

Cultural sociology provides insights into ways of rethinking the learning of science. From this perspective, as social life occurs, participants enact culture in fields structured by resources, which are recursively interconnected with participants' power to act, referred to as agency. The structure of a field, such as a science classroom, is dynamic and consists of human (e.g., people and social networks), material (e.g., space, time, and equipment), and symbolic resources (e.g., status, relationships, and qualifications). These resources can be accessed and appropriated by participants as they exercise agency with or without conscious awareness, to produce schema and practices that reproduce and transform the culture of science.

A fundamental theorem of cultural sociology is the dialectical relationship between agency and structure (Sewell, 1992). Agency requires access to the resources of a field and the cultural capital needed to appropriate them; individuals use resources to meet their goals and, in so doing, change schema and practices which become part of the structure of the field and resources for the production and reproduction of culture (i.e., learning).

Within a field, culture is enacted at nodes that are spatially and temporally distributed. In our research we explore patterns of coherence and associated contradictions in the participants' practices, focusing on their interactions with human, material, and symbolic resources. The nodes we have studied include lab investigations, demonstrations, activities involving the chalkboard, seatwork, and whole class and small group discussions. Since these nodes are distributed within a field such as a science classroom, there is considerable overlap in the culture enacted at any one of them, and practices and schema enacted at one node can be resources for the agency of participants anywhere in the field. For example, a teacher's explanation to a group of students at the chalkboard is heard by students involved in small group discussions and can serve as a resource for their practices.

Practices are patterned actions, often enacted in a field without awareness and in accordance with an agency | structure dialectic. When they are enacted, practices become part of the structure of a field and can be appropriated by all participants (i.e., they become resources for the agency of all participants). For example, if a student decides to argue over the interpretation of data, the argument becomes a resource for the student who initiated it and all those who experience it, whether or not they choose to participate overtly. Unless the structure of a field creates resonant conditions for dispositions to enact particular culture, it is possible that other culture will be enacted and a teacher might erroneously conclude that students are unable to participate in expected ways. However, what is absent is not particular culture but a structure to afford it being enacted. The deficit perspective that the students are unable to perform might push interactions into a dead end. A more productive path to follow is to consider what a teacher might do to create resonant conditions for particular student dispositions associated with participation in science (e.g., to argue).

Engeström (1987) uses activity theory to explore social life in terms of historical, cultural, and social constructs. The approach examines activity in ways that interconnect the individual with the collective and explores agency | structure relationships in terms of the objects and the activity, the resources to support and constrain agency, and the social and political agreements that sanction practices that support progress toward collective and individual goals. Key foci for activity theory are patterns and contradictions that arise in analyses within and between activity systems. As in cultural sociology, contradictions are not seen as sources of error, but a normal part of social life, to be understood in relation to patterns of coherence. A key component of transforming activity systems is to understand and remove contradictions (either by strengthening or eliminating them). Activity theory offers some advantages for analysis because of the way in which agency, structure, and practices are packaged and the emphasis within an activity system on the creation of collective understandings and practices. Historical, cultural, and social events and phenomena usually are analyzed in terms of nodes, commonly represented as the subject, object, tools (including language), rules, division of labor, and community. Analyses are undertaken within and between activity systems at and between each of these nodes with attention being focused on the identification of patterns of coherence and contradictions to those patterns. In our research, we regard activity theory as reducible to cultural sociology and each is used interchangeably as constituents of an overarching theoretical framework.

The sociology of emotions (Collins, 1993) enables us to understand interactions and the extent to which they become resources for creating a community of learners and contribute to a positive learning environment. Collins maintains that successful interactions are associated with positive emotional energy and unsuccessful interactions can lead to a buildup of negative emotional energy. Within a field, participants will tend to engage in interactions that produce positive emotional energy and avoid those that produce negative emotional energy. However, interactions do not occur in isolation from one another, and it is important to consider interactions in relation to those that came before and those that come after (i.e., to consider chains of interactions).

Synchrony addresses the coherence of participants' practices during interactions. For example, as a teacher explains a procedure, a student might nod her head to indicate she is following the oral presentation. Similarly, a humorous anecdote might be followed by smiles and laughs from listeners. Thus, evidence of synchrony might include eye contact, head movement, body orientation and movement, smiles, facial expressions, gestures, verbal utterances, creating and accessing inscriptions, and interacting with human and material resources. Chains of synchronous practices can generate positive emotional energy and lead to entrainment (i.e., chains of interactions that connect and build on one another). In contrast, when asynchrony occurs, an interaction might peter out and fail. An example of asynchrony is when a teacher provides an explanation of a procedure and students shake their heads in frustration. In this instance a buildup of asynchronous practices can produce negative emotional energy and entrainment probably will not occur.

Mutual focus, another attribute of successful interactions, requires participants to establish a common focus for an interaction (i.e., an object) and understand the object from the perspective of others. For example, if a teacher asks a question, a successful interaction necessitates that others hear the question, understand what is being asked, know that an answer is expected, and agree to formulate and volunteer a response. When participants understand one another's perspectives and take them into account, the likelihood of an interaction being successful is increased. Resources for mutual focus from our research include teacher talk, inscriptions on the chalkboard, charts and textbooks, equipment, materials, and verbal interactions among participants.

Entrainment occurs when practices interconnect with, or are anticipated and provide a structure for, subsequent practices (of the actor and/or other participants). Entrainment seems most likely when there is an acceptance of culture, as it is enacted and structured within a field, and is associated with a mutual focus and buildup of synchronous practices. For example, in a lesson on balancing equations, a teacher might write the formulas for reactants and products and show students how to predict products from reactants and balance the equation. Entrainment might involve students exhibiting practices similar to those of their teacher as they balance equations in their notebooks (Olitsky, 2003).

Toward Science Fluency

Transfer, the capacity to enact science fluently in fields away from the school, is a key goal of science education. Being fluent might reflect expanded agency and lead to the appropriation of resources in novel yet useful ways as science knowledge, consisting of practices, facts, concepts, skills, interests, attitudes, and values, is enacted without hesitancy, in timely and appropriate ways. Fluency in a field embraces the metaphor of flow (Csikszentmihalyi, 1990). Discrete actions are coordinated and interwoven with practices to constitute a seamless whole as participants appropriate resources and participate in social life.

Fluency has to do with the flow of practices as participants in a field interact. If fluency is to occur, it seems critical for successful interactions to support the emergence of solidarity among the participants. Roth and his colleagues describe spielraum as practices being enacted without conscious awareness in ways that are anticipatory, just in time, and appropriate (Roth, Lawless, and Masciotra, 2001). This seems most likely to occur when the structural arrangements in a field resonate with dispositions to be enacted, allowing for an activity to flow and involve all participants appropriately. Breaches can occur when practices from one or more participants shut down the practices of others. This can happen in many ways that are important to understand. When a practice is asynchronous with those that preceded it, anticipated resonances do not happen, expected structures do not occur, and flow is breached. This creates an unsuccessful interaction and a catalyst for the production of negative emotional energy, which militates against fluency.

Solidarity occurs within a community when positive emotional energy, mutual focus, synchrony, and entrainment occur and lead to the emergence of a group identity. According to Boykin (1986), there exists a disposition within African American culture for communalism, a tendency to consider participation from the perspective of the group rather than self. If this is so, there may be a disposition toward solidarity within a group of African American youth. It is important for science education to be structured to take advantage of dispositions such as those identified by Boykin. Solidarity implies that participants within a community act together with common purpose and accept others' cultural enactment, including their roles and agency. Solidarity is associated with collectives of various sorts, including shared goals and responsibilities. When solidarity occurs it may be characterized by coparticipation (Schön, 1985; Tobin, 1998), including the presence of incomplete sentences as verbal expressions trail off because head-nodding indicates that participants understand and do not require further explanation, a higher incidence of overlapping speech, and completion of one another's sentences. These examples of coparticipation serve as evidence of entrainment and can be expected to generate positive emotional energy when solidarity exists. Hence, solidarity is a form of symbolic capital that grows from social networks that produce positive emotional energy and coherence in the practices of participants in a field. The emergence of solidarity is gradual and involves the exchange of social, cultural, and symbolic capital (especially respect) during successful interactions.

What seems salient to the teaching and learning of science is that the development of solidarity, as a form of symbolic capital, is a critical step in the development of science fluency in a class. If there is solidarity within a group, then interactions between participants might produce positive emotional energy, synchrony, and entrainment. When solidarity and positive emotional energy are associated with a field, it is possible for activities to be viable foci for a curriculum even when students do not regard them as intrinsically worthwhile or interesting. This was shown convincingly in an example presented by Olitsky (2003) in which students participated in the balancing of equations because of the solidarity that had grown among them and the positive emotional energy associated with coparticipation in whole class activities involving the chalkboard.

In a science classroom, it is important to realize that not all practices can be regarded as science. Since the students in the science classroom are participants in many other fields, they will enact practices from those fields unconsciously.

Accordingly, there needs to be a process of becoming aware of which practices are conducive to science fluency and which are deleterious. In our research, a field in which participants became aware and developed collective understandings and resolutions is cogenerative dialogue (chapter 8), in which a teacher, researcher, and students identified salient issues from shared classroom experiences and cogenerated new roles and associated divisions of labor for subsequent lessons (Roth and Tobin, 2001). Becoming collectively aware is a critical step for teachers and students to deploy cultural capital in ways that are conducive to science fluency and to eliminate practices that do not connect well with science. A priority is for science to be enacted such that students and their teacher are able to coparticipate in culturally adaptive ways and use their cultural capital without being shut down.

Because practices are created within specific fields, they are usually considered to "belong" to the field in which they are produced. So, learning to argue with peers in the streets may result in an effective form of argument, but one that is different than one that is employed by scientists in a research lab, or students in a science classroom. However, fields are weakly bounded and culture belonging to one field can be enacted in others, often to the advantage of the actor. It is important that teachers recognize students' practices that emanate from other fields and can provide a foundation for success in science. Recognition may be conscious, or if teachers have similar cultural and social histories to those of their students, they might accept their practices in stride (i.e., anticipate them), without awareness, and create opportunities for learning. Teachers who do not recognize the learning potential of practices emanating from fields outside of the classroom might shut them down, assuming that they are disruptive to learning and do not belong in a classroom. Shut-down strategies by teachers can be conscious and unconscious and might involve little more than a facial expression, a gesture, or movement of the body. Examples of practices that are often shut down by teachers because they do not belong in the classroom are those associated with students moving around the classroom, social interaction with peers, forms of argument that are loud and seemingly aggressive, and rhythmical tapping and body movements (Elmesky, 2003). Ironically, practices such as these align well with dispositions identified by Boykin (1986) as characteristic of African American culture, incorporating communality, a social time perspective, verve, rhythm, movement, oral expression, and expressive individualism. If shut downs are pervasive, important components of the cultural capital of youth might be suppressed and students would experience symbolic violence, associated with a buildup of negative emotional energy, frustration, and low interest in science.

Teaching is partly rational and mainly a form of praxis that is enacted without awareness, consisting of practices that respond to the unfolding events of a classroom in anticipatory and appropriate ways. For the most part, teachers do not think consciously about what they are going to do in the next moments, but enact teaching practices that are responsive to the field and its dynamic structure. It is only when there is a breach in the flow of activity that a teacher takes stock and decides what to do next (Tobin, 2000). Teachers might plan ahead based on reading, studies, and a value system honed by hours of discussion and reflection. However, when teaching occurs, the enacted practices become part of a dialectical relationship between agency and structure, responsive not only to plans and explicit schema, but also to the students' practices, other components of the field's structure, and practices and dispositions the teacher can enact, but about which she or he is unaware.

In the remainder of this book, including this chapter, the theoretical framework described above is employed in research on the teaching and learning of science in urban high schools. Depending on the foci of the particular chapter, different aspects of the framework are emphasized; however, each chapter contributes to an emerging portrait of understandings that shows how the future for urban science education is recursively interconnected with its past, a past that contains the seeds for trajectories of science fluency that can yield rich harvests. In chapter 17, we adopt a dialectical perspective in looking at the past and future of urban science education; syntheses based on emergent understandings that unfold in sixteen chapters, each grounded in the social and cultural lives of the participants in this research. In addition, the research contained in the following chapters rejects the utility of research *on* others and consists of research *with* others, always taking care to project, respect, and protect the voices of participants. In this spirit, rather than begin this journey with analyses of the teaching of others, we begin with research on my own teaching at City High.

Cultural Miscues and Contradictions

During the first six months of 1999, I cotaught with Mr. Spiegel, the only science teacher in the *Incentive* small learning community at City High. The school bulletin lists *Incentive* as "an academic and resource program to assist students who need to acquire additional academic credits because of extended absences or other extenuating circumstances. These credits will enable the students to achieve appropriate grade level or graduation requirements." Beyond the rhetoric of the bulletin, *Incentive* was a dumping ground for problem students, a place where the most challenging of students were sent, thereby making it easier for faculty in other SLCs to focus their teaching on the remaining students who were more motivated to learn.

When I first began to teach in *Incentive*, many students I spoke to did not respond and seemed to ignore me entirely. This problem was exacerbated by my inability to understand what students said, and I usually had to ask them to repeat their utterances, which was frustrating to them and me. I now understand that it was a source of negative emotional energy for some, and an opportunity for others who seemed to play a game, saying things they knew I would not understand and showing their disrespect of me for failing to understand them. However, I felt that most students regarded my failure to understand them as disrespectful and wanted to see efforts from me to learn their dialect and thereby make sense of what they were saying.

Successful interactions involve exchanges of cultural, social, and symbolic capital. It seems intuitive that, in order to teach science to students, you need to be respected by them and accepted as their teacher. What is not quite so intuitive is what to do to attain and maintain the necessary levels of symbolic capital (e.g., being acknowledged as "my teacher" and earning the respect of students). For most teachers, the basis for successful teaching is the enactment of successful interactions with students, leading to the emergence of status indicators that shape students' identities, including whether or not they are willing to learn from others, and the extent to which they respect one another.

Analyses of a vignette selected from one of my initial lessons in *Incentive* show my use of the chalkboard to establish a focus for the day's activity. In a one minute segment of the tape I listed three tasks on the chalkboard and briefly described each of them, making sure that students knew what was to be done and the autonomy they had in terms of which tasks to complete, who to work with, and what resources to access. I provided them with the option of completing a worksheet on the relevance of science to their neighborhoods, using the Internet to identify chemical and physical properties for a given element (one property for each letter of the alphabet), and doing chemistry labs involving the production of colorful precipitates. I listed the options on the chalkboard and used gestures as I explained what was involved in each option. I spoke slowly, deliberately, at times emphatically, and used long pauses to punctuate bursts of speech. However, my tone was even and there was little evidence of verve, rhythm, or expressive individualism. Similarly, my movement in the class was slow, nondemonstrative, and restricted to the front near the chalkboard. The emotional energy was flat, even though most students were compliant.

The transition from a whole class interactive activity to an individualized activity was seamless and took only two seconds. Most students worked alone, although they were seated in groups of three to four and were able to converse as they participated in their selected activities. I began to circulate around the classroom, greeted each student with "good morning," and assisted him or her to make a selection and get started. Surprisingly, nobody opted to do the lab activities.

I interacted briefly with students at eye level, usually for only one to two seconds and checked whether they had the tools needed to participate in their selected activity, such as pencils, paper, and computer disks. My interactions with students were softly spoken, conversational, and sincere. My goal was to be helpful and most students were task-oriented and not easily distracted. For example, when I spoke to Deidre about some items on the worksheet, the others at her table did not look up from what they were doing.

Some interactions with students were successful, involved chains of synchronous actions, and generated positive emotional energy. For example, in an interaction with one student, I lowered my head to make eye contact, the verbal interaction was conversational, and the student nodded affirmatively and gave short verbal indications that she followed my explanation (e.g., yes, uh huh). When I walked to the chalkboard, her head and eyes followed me and her body movements were consistent with her attending to my talk and associated gestures. Later, with the same student, I knelt down and spoke at length with her about dangerous gases in the home. During this interaction there was ample evidence of synchronous actions including short periods of eye contact, head nods, the use of affirmative verbal fillers, brief smiles, and efforts from the student to enlist others to join the conversation. Although this form of interaction was successful and highly desirable, it was unusual in the first eight weeks and represented a contradiction to the more familiar pattern. Most interactions were asynchronous, with students breaking eye contact with me, making eye contact with peers, and either speaking with others in opposition to me, laughing out loud, or using facial expressions suggestive of a lack of affiliation with me.

My use of proximity desists catalyzed negative emotional energy. For example, as students engaged in strategies that were potentially disruptive to others, I moved close to them and, although I did not usually address their practices directly, I asked questions or looked at their work in an effort to focus their activity on the assigned tasks. My practices seemed to be oriented toward maintaining control by anticipating and preventing disruptions. Consequently, much of my teaching involved struggles over control instead of interactions to mediate learning.

Playin the Teacher

Although Amirah was twenty minutes late, I let her in when she knocked on the door. I greeted her quietly, handed her a worksheet, and offered to help. She shook her head almost imperceptibly and seemingly ignored me. Deidre laughed out loud, possibly an act of public approval for her friend's refusal of my offer of assistance. In an endeavor to moderate and focus her behavior, I moved closer to Deidre and others at her table. I queried Deidre about the questions she was responding to and asked if she had a carbon monoxide detector installed in her home. Kamica showed interest and interacted with me about carbon monoxide, its poisonous nature, and the fact that it was odorless. After a short interaction, I referred Kamica to a reference book and, because Amirah and her group on the other side of the room had begun to laugh loudly, I moved closer to them. Unwittingly, I was responding to a student who was unsettling me, leaving Kamica who was willing to learn to refocus Amirah who was likely playing a game (see Elmesky, chapter 5, for research on *playin*).

Play rituals frustrated me, but did not disrupt most students. In one playful twenty-one second interlude, Amirah moved to the center of the class and carefully watched Deidre working. She then decisively moved to Deidre's desk and grabbed a spare pencil, quickly retreating back to her own desk. Deidre called out in an exasperated, but playful way, "Gir::l↑!" Amirah gave a chuckle as she reached her chair. Others began to laugh, and Deidre remarked "Gonna hurt yu!" Amirah rapidly responded "I'm gonna hurt you::." Deidre then gave a cry of frustration (Aaargh!), and engaged in brisk repartee with Amirah. As quickly as it started, it was over. The two protagonists smiled at one another as their eyes met and they resumed their work. Instead of continuing to teach, I moved closer to Deidre (possibly in an attempt to control any reaction she might have), with no apparent effect, since she ignored my presence.

Learning by Teaching These Students

Eight weeks into my teaching in *Incentive*, there was evidence of significant changes. I still emphasized labs and associated demonstrations as well as activities of short duration. However, my movement around the class was much quicker and purposeful. I tended to ignore students who were inattentive or those who were potentially disruptive, and my speech was noticeably faster. Also, there was evidence of my own expressive individualism and as I taught I made light-hearted quips (e.g., Charles is having fun!) and used my cultural capital in the form of Australian colloquialisms (e.g., Watch out mate!). Even though I was much quicker and briefer, there still existed a tendency for me to talk too much and not to build on instances of positive emotional energy. My interactions with students who were not participating were lively and short (e.g., Let's get busy!) and, with few exceptions, I did not get "in their face." Only occasionally did I continue my earlier tendency to annoy students who were not having good days. Instead, I often backed off and allowed them to get involved on their own terms.

As I moved around the class, there was significant evidence of solidarity and a build up of positive emotional energy as students worked together and interacted with me when I joined their group. I ignored potentially disruptive students and, for the most part, so did their peers. Disruptive practices did not lead to entrainment, and students chastised their peers when were annoying (e.g., "Make sure those washers do not come flying off over here"). My interactions with students were brief and fast-paced and I minimized a tendency to be repetitive and labor the point. Students showed entrainment and solidarity by letting me know when they achieved success (e.g., by calling out "I got twelve! Heh, where'd dude go? Heh, I got twelve.").

As necessary, I quickly dealt with rule violations, such as the use of obscenity, and swiftly resumed teaching as a sign to the students concerned that they were welcome to participate, despite the rule infraction. For example, on one occasion my use of a smile and voice intonation while addressing a student who had cursed minimized a buildup of negative emotional energy. Then, a quick return to doing science created possibilities for "moving on" and for building positive emotional energy—by not focusing on what might just have been a student's slip of the tongue. Later in the lesson, the wisdom of following this path was evident when the female who had made an obscene remark took the lead in conducting an experiment, making sense of data, and sharing what she had learned with others.

Becoming Like the Other

Initially my whole class teaching was slow and deliberate; mainly a strategy designed to make sure that students would understand my Australian dialect. Although I always kept whole class activities short, the pace of delivery within utterances contained many pauses and was a reflection of my years of research on wait time and practice at using a long wait time in all of my teaching since the late 1970s (mainly at the college level). What I did not realize is that successful interactions with these students were more likely to be built around lively and fluent oral presentations and incorporate characteristics that would earn the respect of African American youth and afford dispositions to act in synchrony, for example, to employ variations in rhythm, rhyme, pitch, and timbre. Broadly speaking, the changes in my teaching reflected the adoption of dispositions more harmonious with those attributed to African American culture by Boykin (1986). Videotapes over this time reveal that my teaching showed more body movement, greater verve, and an expressive individualism that incorporated a faster and more fluent style of speaking, and employed greater tonal variation. These changes were not effected consciously and occurred by my teaching [with] these students for an extended time.

Culturally Adaptive Teaching

After a semester of coteaching with Spiegel in *Incentive*, he was invited to teach in a different SLC, where I cotaught with him for a year, until he left to teach in a suburban school in another state. When Carambo (chapter 9) replaced Spiegel I cotaught and undertook research with him. In fall 2002, a major restructuring of CHS resulted in a reduction in the number of SLCs from nine to six. Carambo became the senior science teacher in the *Science, Engineering and Mathematics* (*SEM*) SLC, for students seeking to go to college. I continued to coteach with Carambo and the remainder of this chapter involves an analysis of my teaching in *SEM*.

The students from the *SEM* SLC were much more oriented toward academic learning than those in *Incentive* and most of them wanted to pursue college-level studies. My coteaching (with Carambo) of these students showed that they accepted my position as a teacher and were courteous in their interactions with me. When I taught, they were attentive and I was able to teach in a conversational manner. With these students there was much more evidence of the joy of learning and positive emotional energy. The students were persistent and did not lose attention when explanations were lengthy. Even so, my pace of delivery was fast and I shared the talk between as many participants as I could. I did not belabor my explanations, relying instead on the students to make their cases and argue over different perspectives. In comparison to my experiences in the other SLCs at City High, there was more student talk in my presence, and also signs that the students assumed responsibility for the success of small group discussions. For instance, in a discussion of density, it was notable that the positive emotional energy of the discussion increased appreciably when we began to discuss the heating of syrup, an example one of the students had volunteered. Throughout several vignettes there was evidence of synchrony, in the laughter and smiles, eye contact, leaning forward and backward, and chains of verbal interaction in which successive speakers took into account what was said previously. Hence, there was evidence of a mutual focus that was maintained by most of the participants, not just the teacher.

The following analyses of my teaching of density to a group of nine students provide insights into the emotional energy that was engendered during the lesson. The vignette focuses on me assisting students to understand how temperature affects the density of a substance. The following transcript contains interactions that occurred after an explanation from me and involves Tensie and Marissa as the main protagonists. Tensie argues with Marissa about whether or not a heated object changes its mass. During the brief interaction, students present opposing positions and at times raise their voices. There is considerable evidence of positive emotional energy throughout the interaction and no evidence that negative emotional energy is generated either by the argument, the presentation of opposing views, or the public nature of the interaction. The pace of the verbal interaction is brisk, with short pauses between speakers, and there are examples of Tensie's speech trailing off as Marissa makes her case. Turn-taking is associated with head nods, eye contact, smiles, and changing intonation to project a pleasant tone. Throughout the interaction there is synchrony in the actions of the participants, including me. Signs of attentive listening are evident in my eye contact, body orientation, and head nods. Similarly, the other students in the group also participate in a variety of ways that project synchrony and a positive emotional tone.

Speaker	Time	Text
Ken	00:00:00	As the volume got bigger we've experienced that but the question is what happens to the mass? Some people say that it gets lighter. So that would mean the mass gets less. I'm gonna say the mass [stays the same. (*Tensie seems to be cued to speak by my rapid head shakes*)
Tensie	00:13:13	[Shouldn't it stay the same?
Marissa	00:15:11	Why would it stay . . . the mass stay the same?
Tensie	00:16:00	=It would stay the same because it's still the same weight but it just got thinner. It's still the same amount of syrup but it just got thinner.
Marissa	00:23:18	No. But you could have some ice right?
Tensie	00:26:00	Remember like when we did our
Marissa	00:28:00	=It got bigger. It became bigger. The weight get bigger.
Tensie	00:32:00	The mass. If you weigh. If you weigh, uhm:: ah, that juice. If you weigh that juice, whatever, it is . . . it's gonna stay the same unless you drink some.
Marissa	00:39:00	=No:: Because if you freeze it and then it thaws out it's

		gonna have extra weight when it becomes water.
Donna	00:45:28	=yes.
Marissa	00:46:10	(*Smiles*) That's water (*nearly inaudible*)
Tensie	00:48:05	Why::?
Marissa	00:48:27	Because, it gains weight.
Tensie	00:50:16	It's gonna stay the same cuz it's just thinner like you said like syrup. If you weigh that syrup and all right say it's twenty five grams and then you heat it up and it get thinner it's still gonna be twenty five grams unless it uhm::
Marissa	01:05	=sorry. I get you (01:06:00)

Initially, Tensie endeavors to establish a mutual focus by discussing a lab they performed the previous week. Marissa enjoins the effort and interprets the results to support her understanding of heating a substance and its change in mass. Tensie then changes focus, discussing instead a bottle of juice that is in front of Marissa. Again Marissa accepts the change of focus and discusses the issue of density in relation to the juice. Over time, the mutual focus and synchrony provide a strong pattern of entrainment that allows Tensie to persuade Marissa she is incorrect. The vignette also provides evidence of solidarity within the group. The speakers listen attentively, and argue persuasively, while the others appear to accept roles that recognize the value of argument in learning. There are no efforts by others to take the floor, and ample evidence that they follow the flow of the verbal exchange.

Start	**End**	**Oral Text**	**Marissa**	**Tobin**
00:00	02:28	As the volume got bigger we've experienced that	*Head nods* (0:07-1:06) *Uh hu* (1:06-1:24)	*Head/body bobs* (to 2:06) *Looking deep into group* *Eye contact with Marissa* (2:15)
3:16				*Break eye contact, look to my right, toward board*
3:23	6:09	but the question is what happens to the mass?	*Yes* (3:04) *Right arm moves rhythmically* *Looks at density formula* (3:29) *Maintains eye gaze on board*	*Look at density formula* (3:29) *Points to inscription* (4:25) *Looks at group as a whole*
6:13			*Watches me talk*	*Rotate right arm away from board and rotate*

				body toward group
6:16	8:15	Some people say that it gets lighter.	*Has eye contact with me* (7:14)	*Rotate body to face group. Both arms gesture in front of me, fingers pointed upward. Eye contact with Marissa* (7:14) *Emphatic gestures*
8:24			*Break eye contact, look left*	*Break eye contact, look right*

Microanalyses of the opening comment, which took about thirteen seconds, show that the oral component of the interaction is supported by a variety of non-verbal exchanges that set the stage for a successful chain of interactions. The above analysis involves the first eight seconds of the vignette and focuses on Marissa and me since we were close to one another and had eye contact during this time. Marissa synchronizes her actions with my verbal and nonverbal actions, and I accept overlapping speech and approve of students having the role of coteachers. At 8:24 Marissa appears to realize that she and I have different ideas about heating an object and its change in mass.

The final transcript, shown below, contains less than six seconds of interaction, during which Marissa shows that she is rethinking her position; creating a space for Tensie to become involved. Marissa breaks eye contact with me, and enacts synchronous actions that suggest she is reflecting on the evidence (moistening lips, movement of eyes sideways and then upwards) as my statement contradicts her understandings. It seems as if this moment of disagreement set a context for the interaction that was to follow and did not engender any negative emotional energy. Instead the interaction chain between Marissa and me contains synchronous interactions that point to a mutual focus and anticipates the involvement of others in a discussion that was about to begin.

Start	End	Oral Text	Marissa	Tobin
8:25	10:22	So that would mean the mass gets less.	*Rotates body left* (9:08) *Moistens lips* (9:09-9:14) *Appears to say yes*	*Eyes sweep across whole group*
10:24			*Eye movement to right and up*	*Look toward Tensie*
11:10	14:09	I'm gonna say the mass stays the same.	*Blinks eyes* (11:13) *Full rotation to Tensie who begins to speak at 13:23*	*After saying "I'm," I begin to shake my head to show disagreement and enact four complete cycles as I state*

"gonna say the mass"
Overlapping speech
with Tensie. As she
speaks, I say "stays the
same" and my head
nods affirmatively.

My teaching about density arose from queries from students about the relationship between temperature and density. The students wanted to better understand the relationship and clustered around me as I began to teach. That considered, several salient points warrant attention. First, the interaction was integrally associated with canonical science that was planned a priori to be coherent with city, state, and national standards. Second, I established a focus for an interaction chain that flowed seamlessly from me to Tensie and Marissa, and through nonverbal interactions, included all nine students and me. Third, even though Marissa disagreed with Tensie, she listened to her attentively and accepted shifts in focus without breaching flow. Fourth, Tensie maintained the object of the sequence of interactions, to understand relationships between temperature and density, but established foci that she expected Marissa and others to accept, regard as relevant, and perceive as interesting. Fifth, all students and I were attentive to the interaction chain, appeared to benefit from the outcomes, and in our nonverbal participation, showed signs of synchrony and positive emotional energy.

Looking Ahead

Over a period of about seven years at City High I have seen three different principals and several reorganizations of the SLC structure. These changes have been motivated by the well-intentioned efforts of hard-working professionals, many of whom have been involved in our research. It took me some time before I began to see the brilliant talents of over two thousand African American youth, a wellspring of unlimited potential. The key questions for those of us who seek to educate these youth to reach their full potential is how to identify the capital they bring to the school and provide a curriculum that can build on those resources to craft canonical ways of knowing and begin to mobilize new social trajectories and bold horizons. As we gain familiarity with the students and their worlds, we are also challenged to attempt to widen the canon, that is, expand our notions of how science can be done, what science participation looks like, and what can be considered science.

My deficit perspectives faded because of the experiences of being with these students as their teacher. Their ways of being and knowing were foreign to me and I struggled to earn their respect and the status of being their teacher. However, in the process of this struggle, I sought and utilized new ways of looking at what was happening, what was possible, and how to construct and enact culturally adaptive curricula.

During my time at City High I have learned that a critical barometer of successful learning environments is emotional energy. As a teacher researcher, I learned from this longitudinal study that teaching practices can be adapted to avoid negative emotional energy since it is a poison that can kill off interactions and the emergence of solidarity. Conversely, instances of positive emotional energy can be contagious and provide foundations for chaining successful interactions, synchrony, entrainment, and solidarity among participants.

Becoming culturally adaptive necessitated radical changes in my teaching; some were conscious and others were beyond my awareness. Although it was essential that I learned about the centrality of respect, akin to a currency for these African American youth, it was the role of disrespect that was most salient. My teaching had to be changed to reduce the opportunities students had to show their disrespect for me. The use of student researchers, as teacher educators, was invaluable. Ed Walls's advice to "back off" and "teach those who wanna learn," sensitized me to an overwhelming tendency to focus my teaching on controlling students who did not want to learn (Tobin, Seiler and Walls, 1999); just as Vin Walsh had taught me so long ago. There were three important consequences. First, I had little time to mediate the learning of students who wanted to learn. This was a source of frustration and negative emotional energy to the most motivated students and prevented me from building on the positive emotional energy that arose from successful interactions with them. Second, my reactive practices provided a structure that some students used in a game-like manner that prevented me even more from teaching science. Third, my proximity desists and anticipatory practice of shutting down distractions were irritating to the students I attempted to control. What is ironic is that student practices that were distracting to me seemed to be self-regulatory when they were left to run their course. Furthermore, they did not unduly distract the students, and those who were playful tended to play and then return to their work without a significant loss of time.

Perhaps the greatest changes to my teaching were dispositional. By teaching the students at City High for an extended period of time, I adjusted my teaching to synchronize better with the students' practices and dispositions. For example, my teaching became faster paced and more energetic, and in so doing, it was more accommodating to the students' dispositions. As more interactions were successful, the results were synergistic and evidence of solidarity emerged.

Editors' Perspectives

The theoretical frameworks laid out in this chapter are utilized throughout the remainder of the book. Our applications of these frameworks in our research on urban science education provide fresh insights that lead to research outcomes that have implications for research, teaching, teacher education, and policy. For example, regarding teaching as cultural enactment draws attention to teaching as praxis, consisting of conscious and unconscious practices and schema that are dialectically interconnected. Analogously, when we consider science as culture, the practices and schema are regarded as dialectically interconnected and the fluency of enactment becomes central in the fields in which production occurs and also in fresh fields, thereby raising the possibility of new ways to study the socially transformative potential of science education.

The realization that emotional energy is a barometer for the success of inter-actions provides signs for participants to monitor interactions as they unfold. The opportunity exists for teachers and students to avoid or shut down interactions that produce negative emotional energy and foster or sustain interactions that produce positive emotional energy. Rather than judging the success of teaching in terms of indicators that value control of a teacher over students, a focus on emotional energy can lead to successful interaction chains, solidarity among the participants in a science class, and an increased likelihood that science fluency will emerge.

Chapter 3

Painting the Landscape: Urban Schools and Urban Classrooms

Gale Seiler and Lacie Butler

In the late 1960s, plans for City High grew out of the school district's formula for school expansion and severe overcrowding at a nearby high school and several of the surrounding junior high schools, which at that time included grade 9. However, as planning progressed, the university and business communities clamored for the new school to be designated as a science and math magnet high school serving selected students. With close proximity to several universities and science and medical research facilities, it was deemed a perfect location. The entire upper floor (referred to as the third floor) of the four-level structure was devoted to science. The labs were large and state-of-the-art for that time. Each lab had a section for classroom desks as well as a section with lab stations. They had gas, water, fume hoods, environmental chambers, incubators, distillation apparatuses, chemical showers, eyewash stations, and lab tables with attached stools. A walk down the halls today reveals a room for the use and storage of radioactive materials, a greenhouse, and numerous chemical storerooms. Mirroring ideas in education in the late 1960s, classrooms were separated by movable partitions, designed to create more open and shared spaces. Some of the third-floor rooms open onto a walkway that overlooks a two-floor atrium.

The intention to make City High a science and math magnet school that would draw talented students from throughout the city and even from suburban areas was not supported by everyone. Building the school meant that several square blocks of rowhouses had to be demolished and their residents displaced. A community group came together to oppose the plan that would destroy a portion of the neighborhood to make room for the high school and other university and science-oriented buildings that were also to be constructed at that time. The members were also concerned with the poor education and severe overcrowding that their youth were experiencing, and they wanted a school that would address these neighborhood problems. Though not able to halt the construction of the high school, the group did manage to block the special admission designation of City High. A compromise was reached in which 75 percent of the students would come from the surrounding areas of West Philadelphia, and the remaining 25 percent would be admitted citywide to the science and math magnet portion of the school. The community successfully fought the opening of a magnet school that they perceived as elitist and doing little to help their community and their children, but the school opened in 1972 on many troubled and conflicted notes.

City High was immediately confronted with major upheavals in terms of teacher strikes and gang violence. The teacher strike of 1972-1973 was particularly acrimonious. City High was populated with students considered as "undesirable," as overcrowded schools in the neighborhood took advantage of the collapse of the original magnet concept and confusion over the attendance zone to offload their least desirable students. Gang violence in the school was almost an immediate consequence, due in part to this "transfer" of the most difficult students to City High, thus mixing students from different neighborhoods in the school. The violence that was prevalent at City High dictated how the school was run. School rules were made and policy changes enacted in response to acts of violence. Understandably, fiscal resources were directed towards the creation of a safe school, and since security costs were high, the instructional budget suffered. From an historical vantage point it seems as if more emphasis was directed to controlling gang activity than to planning curricula. The impact this had on the education students could receive was dramatic, especially with regard to science teaching. Teachers were reluctant to engage students in lab activities since students with chemicals, scalpels, scissors, and lab equipment were considered dangerous. This notion persisted across decades. In 2000, two student teachers were the first in recent memory to have students carry out dissections in a biology class. Teachers taught sporadically, due to constant interruptions from students and administration. Lunchtime was one of the most dangerous times in the day, and for a period of time, school was dismissed early so that lunch would no longer be necessary. Later, rosters were juggled to ensure that students were scheduled for lunch throughout the day. A few very difficult students dictated the overall atmosphere of City High. The climate was far from educational and rules became oppressive.

In the early 1990s, City High was still in turmoil. An article in the *Philadelphia Inquirer* described the commons, the indoor atrium in the center of the building, as a snake pit. Finally, the turning point arrived in 1995 when charters were implemented. Charters brought order to City High. More than two thousand students were divided into six thematic charters that were distributed to various parts of the building and students attended all of their classes within their charter. One of these was a selective admission charter that evolved from the original science and math magnet portion of the school. Charters are now called Small Learning Communities (SLCs). The number expanded to ten at one point, and was reduced back to six in 2002-2003.

SLC Structure at City High

The idea behind SLCs was to dismantle the large, anonymous urban high school. In the mid-1990s, as part of a funded initiative, it was mandated that all high schools in Philadelphia be divided into SLCs. Although many challenges were resolved by the creation of these "schools within schools," many of the same problems have resurfaced and new issues have arisen that contribute to the educational failures still experienced in neighborhood schools like City High. The organizational scheme for each school was intended to shift from a subject and department focus to a setting of caring, support, and personalization. Although the division of the high schools into SLCs reportedly reduced school violence and made them more hospitable places for students, it also resulted in fragmentation of departments, along with their human and material resources. The implications of the SLC structure for science teachers and students repeatedly arise throughout the subsequent chapters.

When we began our work at City High in 1998, the school was divided into nine SLCs. Each SLC, headed by a coordinator, consisted of one hundred seventy five to two hundred twenty five students in grades nine through twelve who were taught all of their classes by five to seven teachers. The SLCs represented a tacit tracking system ranging from two SLCs on the highest floor (with academically more successful students) down to one SLC on the main floor near the office (with the most "disruptive" students) and one in the basement (with academically less successful students). Each SLC was focused around a theme and was designed to foster an atmosphere of community often lacking in a large school. The students purportedly were able to select their SLC, but in actuality their choice was often limited or blocked by administrative policy, enrollment limits, scheduling, and student behavior and academic performance. The number of SLCs at City High has fluctuated over the years as successive reorganizations have occurred.

The spatial segregation of SLCs at City High had unanticipated repercussions particularly in science. Each of the science teachers in the two SLCs on the third floor taught in one of the lab rooms. Because students from SLCs on the other floors would no longer travel to the third floor for science, the remaining labs were not used and fell into disrepair. The radiation room came to be a storage closet and the chemical storerooms were untended. Prior to the advent of SLCs, the school had been organized in a traditional manner along department lines and a science department chairperson led the science teachers. When SLCs were instituted, traditional departments were eliminated. A quasi-formal science teachers' group formed and was coordinated by two science teachers from the third floor. Unfortunately, supplies were not ordered based on consultation among all science teachers as to their needs and wants. Instead, science supplies and equipment were dispersed throughout the school with most of them becoming the property of the two SLCs on the third floor, leaving the other science teachers with access to few material resources. Due to the lack of communication between SLCs, sharing science materials was difficult, and teachers were often known to hoard materials rather than store them in a central location accessible to all teachers.

At the beginning of the 2002-2003 school year, Butler's room contained a set of twenty-seven microscopes. However in early December the entire set disappeared from the room. After several days without return, it was learned that the microscopes had been removed by a teacher who had been teaching at City High for a number of years during which he had come to possess a large quantity of the lab equipment within the school. This was a common occurrence at City High, particularly due to the presence of SLCs, as they created a structure that encouraged the tracking of resources and students (Martin, 2002). For example, a male science teacher on the second floor had collected a wide range of basic science equipment that he stored in locked cupboards in his classroom. His equipment included balances, meter sticks, and glassware. He had enough social capital with one of the science teachers on the upper floor who controlled access to science equipment so that he was able to use additional equipment such as hotplates when needed. In contrast, a female science teacher in another SLC on the second floor had no science equipment available in her room, a regular classroom located in a section of the school building that was distant from the original science laboratories. Occasionally she performed demonstrations, but made little effort to gain materials or to use a lab on the third floor. Hence her classes did not do hands-on activities or experiments, except when a student teacher was in her class. Some teachers might have been willing to share, but it was rarely necessary since the lack of departmental structure insured that other teachers were unaware of who had equipment that they might want to borrow.

Material Resources

Urban schools have been identified by many authors as under-resourced, having few or outdated textbooks and supplies, restricted opportunities for field trips and enrichment activities, inadequate computer and Internet access, and buildings in need of repair. Philadelphia schools are not immune to the widespread nature of what Kozol (1995) has called socially created injuries to intellect, as the following examples from City High will illustrate.

In the fall of 2000, the news media carried stories about the high levels of lead in water in drinking fountains and other sources in the Philadelphia Public Schools. The *Philadelphia Inquirer* listed those schools with the highest levels and City High was among them. In December 2000, signs were posted over the bathroom sinks stating that it was unsafe to drink the water. However, no warnings were placed over the water fountains and no one was sure of whether they were safe or not. Although the city's public response to the schools stated that bottled water would be provided to the affected schools, no bottled water was ever made available. The lack of available drinking water was a serious issue in a school building where the heating and cooling system often malfunctions and the temperature in some of the windowless classrooms can become oppressive. It was also something that troubled some of the students at the high school, as illustrated by a student's question in chapter 6 about the health effects of lead in their school water. As far as we know, this situation was never resolved, though the "Don't drink the water" signs have disappeared. This exemplifies the sort of environmental violence that children are exposed to in their inner city schools and communities on a daily basis.

The computer room in one SLC at City High housed about eighteen computers, ten of which were online. Yet, access to the computers and the Internet became more difficult when building conditions forced an English teacher to permanently move her class into the computer room. The classroom where the English classes were held became so cold in the winter that the class had to be moved to the computer room, thereby preventing the use of the room or the computers by other teachers and their classes; computer access for the entire SLC was limited.

Without a strong tax base, schools in Philadelphia, as in other urban centers, have been chronically underfunded, making the material resources available for teaching generally lower than at most suburban public schools. In nearby Gilford, a high-performing, suburban school district, the per-student expenditure is $19,000, while in Philadelphia this expenditure is $9,000. Even at Philadelphia's Urban Magnet, the top-performing high school in the state, its science labs are a far cry from the facilities at comparably performing suburban schools. The science laboratories are old and have not been remodeled or refurbished. Although the teachers describe them as reasonably equipped with laboratory tables and benches, running water, and scientific equipment, no science classroom at Urban Magnet is equipped with state-of-the-art laboratory equipment for the new century. However, Urban Magnet science teachers report that "resources are plentiful" and teachers "can order any materials they need." This is in contrast to the neighborhood high schools where the division of the large schools into SLCs has impacted access to science rooms and equipment. The decentralized organization and unclear, inefficient procedures for the maintenance and procurement of equipment at these schools has resulted in problems. Though most severe at City High, the SLC structure at Southeast High also lowered the availability of material resources and is apparent in other chapters in this volume (see Otieno and Milne, chapter 12).

While charter schools draw the same level of funding per pupil from the school district as neighborhood and magnet schools, they are able to significantly supplement this fiscal base through additional financial support from corporate and individual sources. Just two years old, Charter High has purchased and completely remodeled a four-story building located in the center of Philadelphia. The science teachers have been able to purchase considerable supplies and equipment to stock their classrooms. The school's largest investment has been in computer technology. Wireless Internet access is available throughout the entire building, and every classroom is ringed with computers, one per student in most cases. Charter High is exceptional in terms of material resources for science teaching and learning, and also in the direct access that science teachers have to the use of financial resources to meet their needs.

In comparison, the building that houses Urban Magnet is very small for nearly one thousand two hundred students, and class sizes range from twenty-five to thirty-three. The middle and high school (grades 5 through 12) share the small dining room, gym, and library. There are two computer labs with about twenty-five to thirty computers in each—one is for the middle school and the other for the high school. In addition to crowded classrooms, the shortage of space means that teachers have to "float" from room to room. It is usually the new teachers who float, and this is particularly problematic for science teaching. New science teachers often have to teach science in rooms without running water or any lab facilities at all. Linda Loman, the author of chapter 14, described it this way: "My first year I had one class held in the lunchroom during first period, one in the library held during fourth, and my Biology class met in the art room two days a week."

Human Resources Overcome Material Constraints

The structural constraints at the neighborhood comprehensive school are great, and in the case of City High they stem from both its conflicted origins and subsequent division into SLCs. However there are numerous instances when human resources, in the form of teacher resource-fullness, overcome historical and material issues to create significant learning opportunities for students.

Classroom Space and Organization

Lacie Butler, a coauthor of this chapter, taught an advanced biology course in one of the science labs on the third floor of City High. This class took place in a large double room in which a central divider was permanently opened. Four doors, two of which were constantly locked, opened into the windowless classroom from each of the four corners. The remaining two doors were locked whenever the classroom was empty, to prevent students from using the room as a shortcut through the halls. In addition to storage cupboards, two environmental chambers were present in the room that could incubate organisms at low or high temperatures, but these were not functioning. The room was equipped with six sets of immovable lab tables with attached stools, and two immovable demonstration tables. Several of the fixed stools had been broken and replaced by chairs. The lab tables were designed to provide access to sinks, natural gas, and electricity. In the initial design, the lab tables were located in the back half of the room, with space for desks toward the front demonstration table and chalkboards, however, the desks were gone, presumably removed to be used elsewhere in the school. Thus the students sat at the lab tables for the entire ninety-minute class period, regardless of the nature of the work, an uncomfortable task considering the hard metal stools. The lack of desks necessitated that the coteachers relocate the "front of the room" to the lab portion of the room. However, rearranging the classroom in this way meant that the chalkboards were located in the other portion of the room. In response to a request for an additional chalkboard or white board, a three- by six-foot vertical chalkboard on wheels was provided and this was placed in the middle aisle between the lab tables.

Had multifunctional factors such as desks, or movable lab tables or stools been included in the physical structure of the room, the teacher would have been more able to better adapt the physical environment as a resource for teaching and learning. As it was, the new arrangement of the room placed everyone in close proximity, allowing for increased interactions between the students and the teacher, yet it was still difficult for all students to see the small, portable chalkboard. For the students, this resulted in competition for prime seats whenever information was placed on the board and effectively decreased the learning potential for those who did not succeed in attaining a front seat. For Butler this necessitated a choice between accepting the structure as it was and sacrificing valuable goals or finding ways to create a more appropriate structure to support student learning.

She decided to decrease reliance on the chalkboard, and began to provide students with a note packet for each lesson. The note packet contained a skeletal outline of the notes for the day (which students would complete) and often diagrams scanned from the text. The inclusion of diagrams was necessary since an overhead projection unit was not available. The use of the note packets reconfigured the possibilities for interactions among participants, providing new opportunities for students to participate and learn. Rather than relying on the chalkboard as the fundamental means of presenting information, the chalkboard became a node for creating dialogue and understanding, as its function became one of conceptual expansion, through the identification of key words, challenge questions, and diagramming. This was in contrast to the original, more traditional use of a chalkboard, as the prime vehicle for information delivery. Thus teacher and student classroom actions and practices changed in ways that created greater opportunities for agency and the development of science fluency.

While the use of the note packets created opportunities for dialogue that may otherwise have been missed, it is important to note that not all structural challenges can be overcome by the agency of the teacher. Although the use of note packets allowed desirable interaction to occur despite the physical structure of the room, the chalkboard still was needed to emphasize and elaborate upon certain points through visual cues and gestures, which are necessary for developing scientific discourse (Roth and Welzel, 2001). As such, whenever the chalkboard was used during the class (e.g., for diagramming, emphasizing, providing instructions, imparting information, etc.) students would compete for positions within the room to gain the best visual access to the chalkboard.

Seiler (2001) conducted a science lunch group on the lower level of City High, in a very large all-purpose room that had three distinct areas in it. In one area there were a number of classroom desks and a blackboard. In another portion there were four tall, round tables with stools that were remnants of days when vocational courses such as electrical shop were offered in the school. These units were bolted to the floor and had electrical wires protruding from them. Each table had a large opening in the center that had come to be used as trash receptacles. The remainder of the room housed a few dilapidated couches, some library cubicles, odd chairs, bookcases, file cabinets, and discarded books. The room was usually a mess. Efforts over the previous semester to straighten it up and get it into shape were in vain. Because the room was used as an extra classroom or to house unsupervised students, it always reverted back to its poor condition quickly. Teachers usually disliked having to teach in this room since its physical setup was not considered conducive to efficient classroom organization. However in this room, the science lunch group was able to create an ethos that was vastly different from the rest of the school. The lunch group members talked and ate lunch together while seated on the old sofas and moved to the vocational tables to carry out science investigations. In this informal atmosphere, authoritative power relations between teacher and students were minimized, and patterns of discourse involved great amounts of student-to-student cross talk (Lemke, 1990). The untraditional nature of this multipurpose room afforded the creation of a community that contrasted with the rest of the school and with traditional science education practices. The unconventional classroom setup, considered a challenge by most teachers and students, afforded the creation of new schema and practices among the science lunch group participants, and contributed positively to the development of agency.

Semiotic Messages

Physical characteristics of science classrooms impact the practical use of space, and they play a role in creating an atmosphere for the learning environment, sending messages to the students about the type of activities to be done and the value to be placed on science (Arzi, 1998). These semiotic messages sent by the structure of a classroom take on added significance because they serve to perpetuate the values and beliefs of the dominant culture (Shapiro, 2000), often becoming embedded in the students' science identities. City High, located in an underfunded urban school district, was thirty years old, and there had been no science department chairperson since the onset of SLCs. As described earlier in the chapter, the science labs had been neglected and misused, and in Butler's room, two sinks would not drain and the gas did not work. While these examples of physical decay may have been totally unrelated to the value that the administration placed on science, they nevertheless sent a message to teachers and students that science was not valued and did not merit functioning and up-to-date facilities.

Butler attempted to create alternative messages for her students. She prominently displayed student work, brought small animals such as fish and frogs into the classroom, and created information centers such as the vocabulary word wall. In these ways she hoped to convey the idea that everyone can participate in science, science can be fun, and science is important. Such changes affected the ways in which students viewed science and their agency within the subject area. When students were asked about the messages the room conveyed about science, their responses first represented the functionality of the classroom:

Giovanni	I feel that the room doesn't make me feel like science is all that important.
Mia	When I look around the room, it looks like we don't really have all the necessary tools in the room to have a good science education.

But later represented a transformed view:

Alicita	It makes science seem fun, like more than just work, but a whole new way of learning!
Lori	Once all the posters and pictures were all hung up, the room reflected our understanding of biology. It made me feel more welcomed also.
Kelli	That people and animals are part of science too.

Lab Equipment

Perhaps the most important form of material resources within a science classroom is the lab equipment. Appropriation, supervision, and maintenance of equipment was complicated by the SLC structure at City High. In preparing for a unit on DNA, Butler keenly felt the truncation of agency experienced by many urban science educators, as her plan for her students to carry out gel electrophoresis was hampered by a lack of materials. Although the electrophoresis chambers and all other equipment were available and functioning, the agarose and DNA samples had not been refilled since the prior use. This situation appeared to be structurally related to the SLC format, the lack of a science department, and budgeting issues. It represented a contradiction between her classroom goals and the organizational structure of City High. Rather than be deterred, she again employed her resourcefulness to use the given structure in creative ways to enhance the DNA unit. Instead of investing time and money in reordering the materials or allowing the structure in which she operated to confine her agency, she opted to engage students in a more immediate and realistic type of activity.

Arriving in class several days later, students found themselves in the midst of a simulated crime scene. Caution tape partitioned off a section of the room containing a tape outline of a body, the book used as the weapon, and evidence samples. Students were given envelopes with their characters' names and roles in the crime, in addition to information on their assigned crime scene investigation team and specific area of expertise. Students formed investigation teams and began rotating through six evidence stations, including footprint analysis, microscopy, fingerprinting, blood typing, testimonial analysis, and DNA fingerprinting. Although students were unable to actually carry out DNA electrophoresis, they were provided with sample results obtained from the Internet and asked to analyze them. While this alternative did not allow students to do electrophoresis, it did allow them to visualize and interpret synthetic results while incorporating their understanding of DNA with additional concepts from class in a real-world application.

Although the students did not learn the procedural details of DNA electrophoresis, they gained an understanding of the concept and its role in forensic science. In addition they gained insight into how scientists compile data from multiple sources in formulating conclusions. To some this may be viewed as a deficient lesson since the students' access to science content was disadvantaged. However, the integration of DNA electrophoresis with microscopy, blood typing, and criminal investigation created a stimulating learning experience that allowed the students to become involved in the process of science. By providing activities that bridge gaps between students' lifeworlds and their classrooms, teachers provide students with a structure that empowers them to take control of learning (Seiler, 2002).

Using their renewed agency, students were able to merge their inquiry skills, cognitive understandings, and affective beliefs about science thus allowing them to integrate scientific literacy with their interests and experiences, something that may not have happened with a basic gel electrophoresis lab activity. In acquiring a strong sense of agency to participate in the classroom, students can make such connections and are more likely to persist in the science field, thereby changing the structure that may confine their future participation in science (Seiler, Tobin, and Sokolic, 2001).

Laboratory Safety

In addition to the availability of functioning equipment and materials, safety issues also created threats to transformative teaching. In Butler's room, a goggle sterilization case was present, but it contained no goggles, as the lab assistant stored the only set in her office. While two safety showers hung prominently from the ceiling, neither functioned and the pull handles had long since been placed beyond reach. And although an eyewash station hung on the wall, there was no water bottle to cleanse the eyes. In addition, the room lacked a fire extinguisher, contained broken glass in cabinet doors, and lacked a doorknob on one of the classroom doors. During a cogenerative dialogue students were quick to identify safety concerns such as the following.

Fatah	There is not enough safety equipment here. What are we supposed to do if there is a fire and we have no fire extinguisher?
Dee	The eyewash station is empty! You can't clean your eyes out.
Trinity	There is mold on the electrical socket. That shower won't work in an emergency, and there's no doorknob. We're trapped in here!

These types of safety concerns have the potential to prevent teachers from performing certain activities within the class. During the unit on DNA, Butler wanted students to extract DNA since she thought it would entice their interest to actually see DNA and to perform a lab viewed by many students as an activity that is only possible for "real" scientists. In addition, it would solidify students' understandings of the containment of DNA within the nucleus of the cell. Typically DNA extraction requires the use of many materials including chloroform, a toxic chemical that must be used in adequate ventilation, while wearing protective lab coats, goggles, and gloves (MSDS, 2001). Not only would it be difficult to obtain many of the needed materials, but it would certainly have been a danger to the students and the teachers to use such a potentially dangerous chemical whose use requires proper ventilation, protective gear, and safety features. But Butler refused to disadvantage the learning of her students due to the structure of the classroom and lack of safety devices. Instead she located a lab procedure that allowed students to extract DNA from strawberries using simple household products. By employing the available human resources (a friend helped her locate the protocol, Explore IT, 2002), not only were the required materials much easier to obtain, but the safety risks were minimal. The resulting lesson allowed for a great amount of student interaction and manipulation of the materials, thereby engaging them as true scientists and creating opportunities to connect science with real life in empowering ways. While the lack of safety equipment in the lab had the potential to prevent student engagement in a valuable lab experience, the presence of an agentic and resourceful teacher transformed the structure and created opportunities for learning.

Material and Human Resources Constrained

Responding to district directives in 1995, the principal of City High disbanded all subject departments and their leadership in order to support the development of SLCs and the appointment of SLC coordinators. Such decisions made in response to material (fiscal) constraints often impact human resources that in turn shape structure and practices and student agency and learning.

Lab Assistant

Although the advent of the SLC structure eliminated a functioning science department and an organization that included a department chairperson, the position of lab assistant remained. While this position could have been empowering for science teachers at the school, it typically had a disempowering effect. During our research at City High a new lab assistant was employed. She was responsible for completing inventory and stocking lab equipment, as well as preparing lab materials as requested by science teachers throughout the building. The lab assistant had no prior science experience, through employment or education. Such a lack of experience greatly hindered her ability to do her job because she did not possess the scientific knowledge necessary for identifying, maintaining, and preparing equipment. This created an access structure that truncated the agency of the science teachers. Often unable to identify the specific equipment requested by teachers, the lab assistant would then have to ask someone to identify it for her. While this could be viewed as a learning opportunity for the lab assistant, her continued inability to identify equipment meant that requesting her services greatly extended the planning and preparation time required by teachers. In essence, the lab assistant created an access structure that complicated the extent to which teachers could prepare for lessons, and diminished the teachers' agency by requiring them to depend on her.

This dilemma was evident in preparing for the forensics lesson that Butler enacted. The inability to locate the lab assistant in the days before the lesson meant that she did not receive the equipment list until the preceding day. The following morning the materials were not prepared as the lab assistant needed help in identifying much of the equipment. While the equipment was eventually obtained and the supplies prepared, the structure created by the presence of a lab assistant served to reduce the agency of the teacher by requiring extra, ineffective steps to obtain materials.

While the presence of a lab assistant at City High had the potential to greatly enhance the science teaching that took place, this potential was not actualized. As a centralizing agent of resources, a lab assistant traditionally can easily locate materials and equipment for teachers, decreasing the time spent in tracking down such resources. Centralizing resources with a competent lab assistant would also ensure that materials were restocked regularly and maintained properly, thereby providing a structure that would free teachers from the arduous search for functioning equipment and allow them more time to plan and enact activities. While part of the inefficiency of the lab assistant structure at City High was due to the lab assistant's inexperience, it was further exaggerated by the structure of the school itself, which lacked a science department. Such a situation demonstrates the need for support from sources external to the classroom, including administrators, other teachers, and the community, to fully transform the unacceptable structures in which urban science educators work.

Without the presence of a science department and science department head, the lab assistant's work remained largely unsupervised. As she had initially displayed great enthusiasm and willingness to learn (Tobin, personal communication), the presence of a department leader could have provided her with encouragement and learning opportunities allowing for her growth in a new profession. In addition, the presence of a direct supervisor may have afforded a higher standard of accountability and professionally appropriate conduct. The lab assistant's practices may have evolved over time to increase her agency, which in turn could serve to enhance the agency of teachers and students by providing them with an effective structure for accessing materials, equipment, and information in a timely fashion. As it was, this human resource fell short of its potential and instead truncated teacher agency.

Science Department Leader

The once-premiere science labs at City High had deteriorated and did not provide an appropriate structure for scientific inquiry. The lab assistant provided little support for teachers and students. Although many factors were responsible for the decline of the science classrooms and procedures, the absence of a science department certainly contributed. The lack of a structure and a person to oversee science matters influenced the science education that took place at City High in additional ways as well. Beyond the physical and material aspects of science education at City High was a lack of cohesiveness. Courses did not follow a logical progression, teachers teaching the same course did not cover the same material, and many teachers taught out of field without a support system.

In large part, the lack of cohesiveness among the science courses was due to the disjointed environment created by the SLC structure where science teachers had little if any interaction with each other. The lack of inter-SLC teacher interaction was related to both the physical and temporal separation of personnel. The schedules in different SLCs were specifically designed so that class periods did not coincide thus avoiding large numbers of students in the hallways at the same time. However the isolation was also self-imposed through the competitive atmosphere that arose between SLCs. Lack of a science department further isolated teachers as there were limited opportunities to create common goals, interact and share reflections, and learn from the experience of in-field colleagues. Without the structure of a science department to create such conditions, it is not surprising that each teacher tended to think only of himself or herself and his or her students, redirecting materials in ways that benefited them.

Lack of Control over Physical Features

Physical factors beyond the control of any of the stakeholders, and thus out of reach of human resourcefulness, also impacted teaching. The temperature in several science classrooms at City High was unpredictable and often uncomfortable, and many rooms had no windows. One day it might be sweltering and the next day it might be frigid, as if the air conditioning was on. Students often complained about the classroom climate and reacted to it in a variety of ways. When cold air blew out of the vents they shivered, shared coats, and asked to go to their locker to get a "hoodie." They often huddled in their winter coats, hats, and even gloves making science activities difficult. When the room was hot they claimed that it was too oppressive to work, that the heat made them sleepy, or that they needed to get a drink of water. This was an additional problem since the water was unsafe to drink due to its high level of lead. Some days the students refused to come into the room because of the temperature and the teacher was forced to relocate to another room. Unclear channels of communication and responsibility made it difficult for teachers to get this heating and cooling problem addressed and temperature extremes continued to elicit uncooperative behavior from the students. Students often expressed anger and frustration with the teachers for their inability to address this problem, an issue that was not within the teachers' control.

The Pivotal Role of Human Resources

When we talk about human capital that can contribute to science classrooms, we mean students and teachers as well as other persons potentially involved with the schools. Human resources of these types can interact dialectically with schema and practices to transform structures, create agency, and alter science pedagogy and participation. However this does not happen uniformly at all the school sites. Though the magnet and neighborhood schools in Philadelphia all suffer from a lack of material resources when compared with schools elsewhere in the state, some schools are richer than others in human resources. Among the schools written about in this volume, important differences exist in the human capital that is available in each setting but the distinctions are not clear-cut.

Teachers

One of the most widely publicized human resource issues facing urban schools is the difficulty of insuring that the classrooms are staffed by certified, qualified teachers (Darling-Hammond, 1999). Though nationwide there are sufficient teachers, the shortage is acute in mathematics and science, particularly in chemistry and physics, and in the urban schools that serve minority students from conditions of poverty (Ingersoll, 1997). Complicated, slow, unwelcoming application procedures and union regulations have historically plagued urban districts and turned away numerous teachers. For example when City High opened, teachers who had been specifically prepared to enact the school's learner-centered curriculum were instead assigned to other schools. Science is particularly prone to out-of-field teaching, since the disciplines of science are often viewed as interchangeable, so teachers certified in biology are frequently assigned to teach chemistry or physics. These sorts of staffing problems are seen to some extent in all of the high schools written about in this volume, though there is significant variation between them.

Human resources in the form of experienced and qualified teachers are more abundant at schools such as Urban Magnet and Charter High. Though teachers at Urban Magnet, particularly new teachers, are often assigned to teach classes outside of their area of certification, these teachers are generally recruited because of their superior teaching abilities and are skilled in the classroom. Due to its charter status, hiring and retention practices at Charter High lie outside the union contract that binds other public schools in Philadelphia, thus freeing the administration to recruit and hire the most qualified and desirable teachers available. However, the small size of the school and staff has resulted in out-of-field science teaching there as well.

The use of human resources is different at Charter High. Teacher participation is valued in making policy decisions and school rules. Teachers and staff are given the opportunity and time to discuss everything from curricula to changes in the school calendar and schedule. While this attempt to privilege teacher voice is an improvement over the typical hierarchical structures apparent in most schools, it requires time and a high level of commitment from the teachers at the school. However, these longer work days and the close-knit community created within the school allow teachers to discuss cross-curricular ideas and more fully coordinate teaching and learning.

Stemming from the division of the large comprehensive neighborhood high schools into SLCs, out-of-field science teaching is particularly common in high schools such as City High and Southeast High. For example, an SLC might have two hundred students and seven teachers, including a single science teacher who would teach all the science that is taught in that SLC, from biology to chemistry to physics. Such is the case with Cristobal Carambo, author of chapter 8 and Anita Abraham, a teacher researcher who collaborates with us (see Elmesky, chapter 5), both of whom are certified to teach chemistry but must also teach courses in all the other areas of science within their SLC. In 2002-2003 this problem was addressed by reducing the number of SLCs at City High to six, thus increasing the number of science teachers in each SLC. Of course this also raised the number of students in each SLC, which impacted the staff's capacity for personalized attention to students.

On Your Own

Teachers who are the sole science teacher in an SLC handle the challenges this presents in surprising ways that are made possible by the absence of departmental supervision and coordination. For example, Campbell, the only science teacher in his SLC has a background in the social sciences and studied little science in his undergraduate degree. He often has the students use a human biology textbook even when the course he is teaching is physical science. He explains that he believes it is important for teens to know about the body. Mato, the science teacher in an SLC on the second floor, is certified in biology and general science but is expected to teach chemistry and physics as well. Usually she just does not teach these subjects, although students' transcripts may say that they have taken a course called "chemistry" or "physics." This autonomy in what is taught in a particular course is possible because there is no departmental oversight and no curriculum review process. The SLC coordinators are usually occupied with administrative and disciplinary issues, and are not knowledgeable in science content.

Spiegel taught in an SLC on the first floor. He was a new teacher with certification in biology although it was not his college major. He often taught life science topics in physical science courses. For example, an engaging activity that was prepared for the biology course would also be done with the students in other science courses such as chemistry or physical science. Spiegel was more comfortable with biology subject matter and more familiar with activities in this area. He was therefore able to do a better job of engaging students in science and motivating them to do the work when he was teaching this subject. Spiegel taught in the *Incentive* SLC (see Tobin, chapter 2), an SLC for students who had been discharged from other SLCs or other schools, or were returning to school after being off-roll or in the juvenile justice system. Though his choice of science topics might be regarded as unprofessional and unethical, Spiegel's actions can also be viewed as agentic in that they enabled him to gain social capital with the students and afford some science involvement and learning.

Teachers responded in different ways to the high degree of independent decision-making and low degree of oversight and accountability with respect to the science curriculum. The lack of expectations of what and how to teach had the potential to be liberating and empowering, as it appeared to be for Butler when, in the face of obstacles, she developed new activities such as the forensics lesson and DNA extraction from strawberries. However other teachers responded by being limited in their appropriation and creation of resources for learning. School and SLC structures interact with teacher agency in a variety of ways to construct multiple approaches to teaching science at City High and these vary in the extent of agency created.

Mato related that she did not take her classes up to the labs on the upper floor because the students did not behave appropriately when they were there and she did not have the preparation time to organize laboratory activities for her classes. She explained that the media production theme of her SLC meant that it was important for students to develop their information technology skills rather than laboratory skills. Unlike some science teachers at the school who accumulate equipment and materials, the resources that Mato stored in her room were primarily print-based. However, there were no computers in Mato's classroom, the computer laboratory associated with this SLC did not have a class set of computers, and usually half of the computers that were there were either not working or had software problems. Thus the development of students' information technology skills related to science was difficult to carry out.

Mato's schema about what the students could achieve and how they could behave, what constituted science, and the need to maintain order, structured science in her classroom. Her science classroom resonated with some of the characteristics identified with pedagogy of poverty (Haberman, 1991), though she appeared to have considerable power to enact science as she saw fit. As long as the students seemed to be working relatively quietly when the principal or SLC coordinator walked into the classroom, the activity in the classroom was not scrutinized. However pedagogical variety and authentic activities were absent from her room. Students were not provided with opportunities to work with materials or connect science to their everyday lives.

Though these inner city teachers are often teaching out-of-field, it is important to remember that there are other less measurable aspects of what teachers bring to teaching than whether they are certified to teach in a certain area, or even whether they do hands-on activities. Although several of the teachers in this volume are among those categorized as out-of-field teachers, they are experienced and capable urban science teachers. In addition they are dedicated to the belief that all children have the right to be taught science in ways that are potentially transformative. Their very involvement in this on-going research serves as a testament to that commitment, and through the research they provide human capital beyond what is written on their teaching certificate.

Students

Information on the socioeconomic status of students is significant in as much as it influences students' access to cultural and social capital that is valued by school and society. The Philadelphia school district is among the ten largest school districts in the country and, while statewide 39 percent of students are from situations of poverty, in Philadelphia 72 percent live in poverty. At many individual schools in Philadelphia the percentage is often greater than 90 percent. At Charter High, approximately 80 percent of the students qualify for free or reduced lunch programs. In Gilford, a nearby suburban school district, the percentage of students from economically disadvantaged homes is only 3 percent. As mentioned previously, the citywide, selective admission schools in Philadelphia serve a much smaller proportion of students in poverty than the neighborhood schools.

While City High and many neighborhood schools in Philadelphia serve segregated students populations, Urban Magnet has a more diverse student population and it is composed of students who have successfully achieved in school. Charter High serves a self-selected population that is more diverse and representative of the city as whole, since it draws students from throughout Philadelphia.

The contradictions tangible in examining the cross-section of schools represented in this book can be seen clearly in looking at human capital and resources at Urban Magnet and how these afford agentic behavior. Urban Magnet sits at the top of the list; it is the highest-achieving school in the city and the state on standardized tests. All of its students graduate and go to college. The school has some advantage over the neighborhood high schools in material resources and economic capital, but is disadvantaged relative to other high-achieving schools statewide. It receives the same funding per pupil from the city, as do the comprehensive high schools and Charter High, but both Urban Magnet and Charter High supplement their financial resources with fund-raising and grants. Lab facilities at Urban Magnet are old and out-of-date but functional and maintained. Short on space, new teachers often teach science in regular classrooms as they float from room to room during the day. At Urban Magnet the science teachers are generally certified to teach some area of science. Since this school is renowned as a high-achieving school, they are able to attract qualified teachers. New teachers are often recruited based on recommendations to the principal. Thus Urban Magnet seems to be somewhat advantaged in terms of its teaching staff.

The greatest difference between Urban Magnet and the neighborhood high schools is in the human capital found in the students and their families. Though the students at Urban Magnet come from throughout the city, only a few come from conditions of true hardship, and these are exceptions. In general the students are from different circumstances and histories than the students who do not find their way to a citywide Urban Magnet. Some students at Urban Magnet come from the same neighborhoods as students at City High, Southeast High, and the other neighborhood high schools, but more frequently they come from neighborhoods that are more working class, or middle class, or white, or diverse. They are generally not from the wide swaths of the city that are nearly all black or all Latino communities where adults are largely underemployed or unemployed. Even if they do come from some of the same neighborhoods as students at the neighborhood high schools, the students are different. These students have acquired social and cultural capital and tools and dispositions that are privileged in school. The selection criteria for Urban Magnet ensure that the students have histories of success in school. They must have all As and Bs in their classes. If their records show patterns of "behavior problems," absenteeism, or lateness, they are not admitted.

The students at Urban Magnet succeed in school and in life because they have capital and resources that afford them success. But we are not saying that students at the neighborhood high schools lack capital and resources or that the capital and resources of the students at Urban Magnet are better than the capital of students at City High and Southeast High. Much of what we have learned from the research in this book counters the idea that Urban Magnet and Charter High have greater human capital. We do not view the students, families and communities of the neighborhood schools as deficient in capital. Rather, the chapters will illustrate that all of the science classrooms we write about, regardless of the type of school and student population, are rich with student and community resources. All of the classrooms have students who bring with them tremendous cultural capital in the form of schema and practices, yet differences exist in the nature of that capital and the extent to which the students' resources are compatible with mainstream schooling. Several chapters illustrate that in the hands of capable teachers, attuned to inner-city teaching, the capital of inner-city youth can be a valuable human and educational resource. Thus the chapters further illustrate the many ways in which practices, resources, and schema interact, and the extent to which they are able to create opportunities for agency as well as contradictions when they do not.

Editors' Perspectives

Public high schools in Philadelphia differ in the ways in which they are organized and in terms of the physical and human resources accessible to them. Frequently, teachers and students overcome structural constraints because they invest personal resources to support and sustain their activities. However, their efforts may fall short because necessary resources are not accessible to them. Although we regard it as laudatory for participants to use their agency to overcome obstacles we recognize the danger of disillusionment and burnout when individuals experience a lack of support for their participation in science education.

For example, it is commendable that teachers spend their personal money to procure materials and supplies, and invest many hours after school and on weekends to maintain equipment and the physical environment. An irony of teaching in urban schools is that frequently budgeted money is not expended and human resources such as lab assistants are used ineffectively. Because policies and practices associated with the school district, the school and SLC all can structure science education, it is imperative that representative stakeholders participate in cogenerative dialogues to produce collective agreements and responsibilities regarding science education and the manner in which it is to be structured. We regard cogenerative dialogues as useful activities for teachers and administrators to identify and resolve contradictions that pertain to science education, either by strengthening or extinguishing them. This is especially important when organizational structures such as SLCs disperse science teachers throughout a school and create challenges for coordination and coherence. Meetings that do not include all stakeholder groups (such as those attended by school and district administrators) can create an illusion of effectiveness but lack the human resources to effect what is collectively decided.

Chapter 4

Organizational Mediation of Urban Science

Wolff-Michael Roth

On this late morning of a sunny fall day, I arrive with a colleague at the main entrance of City High. The building still strikes me as a prison, although I have become familiar with it over the past several years while doing research in this setting. The sense of visiting a prison intensifies as I open the entrance doors and face a metal detector—as always, I step around it but, invited by a coresearcher to sign in, step toward the table where two individuals are engaged in a lively conversation. We walk up and my colleague shows his ID card while saying that I am a visiting researcher.

We stop at the main office to ascertain that our meeting with the assistant principal is still on for later in the afternoon. At the counter that separates the front part of the room, an African American man with a teenager is talking to an individual on the other side. A few other students sit or stand in this area. In the background, on the other side of the counter, several women at their desks talk to one another, and do not acknowledge us. We wait and then catch the attention of the assistant principal, which allows us to set our meeting for four o'clock.

We walk to the stairwell and up to the third floor where the science, engineering, and mathematics (SEM) small learning community (SLC) is located. The assistant principal has told us that Cristobal Carambo, the coordinator of this SLC, is probably in his science classroom, involved in lunch period tutoring sessions. As we approach the designated classroom, we find Carambo speaking in Spanish with two students. One of the two is, as I find out later, a twelfth grade student participating in the tutoring program as a mentor to the second student. Carambo points out two other females in their early twenties, one Asian and the other Caucasian, each working with two African American students. He explains that they are engineering students from the nearby university, helping City High students with science and mathematics.

This science classroom becomes our home base for the remainder of the day, since Carambo, whom I interview later this afternoon, has made this place his headquarters. Those needing him apparently know that they can find him there. In the course of the next three hours, many come and go. The new Spanish teacher in the SLC comes to pick up the materials left by the previous Spanish teacher, who had gone on sick leave. Two students, who have been asked by their science teacher to leave their classroom, Bobby for sleeping (we meet him again later in this chapter) and the other for doing homework for another course, also arrive in the room. Carambo, after briefly talking with them about having a meeting with the teacher later, asks the two to sit down until he has time to ac-company them to "have a chat with the teacher" and resolve the issues of conten-tion. There are also several twelfth grade students apparently returning with class sets of completed PSSA personal information forms, and leaving, after having been instructed to go to another classroom, with a fresh set of forms. All after-noon, Carambo is busy attending to the needs of others. In fact, Carambo had asked the twelfth grade students to assist him in getting the PSSA forms com-pleted, because attending to the needs of others takes considerable time and only allows him to personally complete the task in one tenth grade classroom taught by Juanita Solento, a chemistry and physics teacher in SEM.

When Carambo visits Solento's chemistry class, it begins with a consider-able, twenty-minute delay since the students have been asked to complete the cover sheet for a standardized examination (PSSA) that the school district main office has scheduled for one of the following three days. While they are waiting for the students to complete their task, Carambo talks to Solento repeatedly and at length, both to organize the time for completing the form, and about other is-sues. While moving back and forth from Solento's classroom to his room to get the forms, Carambo interacts with a physics teacher and a coordinator from an-other SLC. Although they evidently want to talk to him, the interactions are icy. I find out that the first teacher had been competing with Carambo for his coordina-tor position; Carambo had "stepped on the second person's toes" while attempt-ing to place a student in her SLC so that he (student) could take the courses of his choice.

At the end of the day, my colleague and I return to the main office. When we arrive at the assistant principal's office, she invites us in although she is currently talking with a school police officer. They are talking about an incident, in which he removed an unruly female student by her arm, and after having been expelled, she had returned to school with a parent, to complain that the police officer had roughed her up. The parent was the one we had seen in the office when we entered the school earlier that day.

Fields and Interactions

During this single day at City High, I spent time in different settings within the school. In each of these settings, I witnessed events and patterns of behavior similar to others I had already seen during previous research stays. But events and behaviors differed between settings—I saw nonteaching assistants and police officers at the entrance, parents interacting with school personnel, students waiting around in the foyer of the main office, science teaching, and tutoring during lunch. Other events were not only new to me but also, as different individuals with whom I interacted in the course of the day told me, were infrequent and even singular. The request to have students fill out the title pages of the standardized test had arrived at the school only that same morning, and there appeared to be a frenzy of activity to get everything set for the next three days, including a change of bell schedule, forms, supervision, etc. The settings were not only different, but also what happened there, most of the time, seemed to be independent from what happened somewhere else. For example, the interactions between the father and the school personnel in the main office had no bearing on the tutoring sessions that we saw immediately after; but he was in the main office because of a series of events following his daughter's actions in a setting other than the office.

In each of the settings within City High, I observed a different set of cultural practices, which, though characteristic, were not enacted in a deterministic fashion. Thus, Carambo's presence during the tutoring session did not influence events in a determinist way but rather contributed to the possibilities. His fluency in Spanish, for example, became a resource to the tutor-tutee relationship. But the interactions with students may not have occurred had I arrived ten minutes earlier and taken up Carambo's time.

To analyze cultural practices within and across multiple settings, the construct of field is useful (Bourdieu and Wacquant, 1992). A field is a system of social and material relations that functions according to its own, characteristic logic (rules), which is partially inscribed in and arises from the sociomaterial resources available to the actors, who themselves bring structure in the form of schema. Classrooms, main offices, teacher staff rooms, and hallways constitute fields, where characteristically enacted practices are related to roles, spatial arrangements, entitlements, and artifacts; all these are resources for actions available to be accessed and appropriated by students, teachers, administrators, and others. Whether they are actually accessed cannot be predicted because of the emergent nature of social action in general.

Any action is therefore doubly structured, by the relevant sociomaterial resources available in the setting and by the schema embodied by the person. However, the structuring processes are nondeterministic, because resources (e.g., an artifact) and schema (the way participants perceive and act toward the artifact) are two, nonidentical aspects of the same irreducible unit. For example, the metal detector at the main entrance is both an opportunity and a constraint for action; different actors perceive it in different ways leading to different actions and the production and reproduction of different aspects of society. Since all students entering the school have to pass through it, this has led in the past to two-hour line-ups before students could get in. Of course, this mediated subsequent actions in the sense that the scheduled lessons including science could not take place as planned. Teachers and students viewed these delays in ways ranging from frustration to relief. Some teachers and students appreciate the presence of the metal detector since it creates a sense of security that would otherwise be lacking for many. My stepping around the metal detector, a manifestation of white privilege, is for me an opportunity not only to show that I find the practice demeaning, but also to question the practice of having the detector there. Both points also show different forms of equality and power being enacted. The students, mostly African American, are subjected to the procedure and have to suffer an effect of power, whereas I, a white professor, can without trouble circumvent it.

Organizational effects are created when different fields interact. Although the cultural practices within the different fields of a school often seem to be independent, they are actually connected by artifacts that move between fields and people that traverse them. Artifacts, including human bodies, structure events (in nondeterministic ways) because they are resources for the actions of people in and constitutive of the field. First, when artifacts move into a different field, they contribute to the structure of events that happen there. For example, when administrators of a school district instruct principals in the system to test all students and all students take the test, many processes and effects take place. A letter or fax containing instructions arrives in a school; the principal uses it as a resource for action, telling the coordinators of the SLCs that testing will take place; she instructs the assistant principal to create a new bell schedule to structure the temporal organization over the next three days. Her actions therefore both produce and reproduce the school as an organization nested in a larger organization. Objects moving between the different settings linked and therefore produced the different levels of the organization. However, the way in which an object structures the events in a setting is not predetermined, and the same form or instruction can give rise to quite different practices as they move across field boundaries.

Second, organizations come about when members, normally from different fields, interact in the same setting. Organizations are produced and reproduced in interactions, as participants use what they understand about the organization as a resource for their action. A teacher who comes to the SLC coordinator because she has difficulties with a particular student crosses from her classroom into the coordinator's office where they seek a solution that may or may not have been achieved in the original field. However, as I show in this chapter, such interactions not only get things done and therefore produce and reproduce the organization, but also contribute to who the various actors are, including their relative position within the organization.

Organizational phenomena require coordination between fields, which requires work because each field is characterized not only by the participants, artifacts, and practices, but also by a particular timescale, duration, sequence, temporal location, deadline, and cycle of events. Time is inscribed within the artifacts, operating routines, organic matter, habitual norms, and sedimented practices (Kavanagh and Araujo, 1995) and social actors are not just subject to time ("there is no time for labs," "students wasted time") but also use time in productive ways ("make time for meetings," "sync our schedules"). Organizations stand and fall with their members' ability to produce and reproduce coordination despite differences in the structures characteristic of each of their many fields.

This perspective on schools as organizational phenomena cuts across the divide often made between micro level, on the one hand, and meso and macro levels, on the other. These levels are treated here as a heuristic for the analysis of a singular but dialectical phenomenon: individual, face-to-face interactions produce the organization, but the organization constitutes a structure that social actors are continuously oriented to. Social actors draw on and use organizational phenomena in their actions, thereby producing and reproducing these phenomena that frame and enable them. Schools, school districts, and society at large are meso and macro level phenomena that are continuously oriented to, produced, and reproduced in micro level social actions and interactions.

To account for the differences in science teaching and learning across school systems and schools, we need a better understanding of organizational mediation of the events that make a school. In the following two sections, I analyze the two processes that have organizational effects, that is, which produce and reproduce an organization (i.e., make it what it is). First, I focus on the effect of boundary objects, material entities that cross from one field into another, on the social practices in the field where they arrive. The name of the concept derives from the fact that boundaries between fields can be recognized when the same object leads to different practices. Second, I show how micro-level interactions produce the school as an organization. Finally, I intimate how boundary objects and micro-level interactions mediate science teaching and learning.

Boundary Objects and Power

The school district office, school main office, and SLCs within the school all are organizationally complex, and they constitute fields or sites for the enactment of characteristic culture. These fields are not independent but are dialectically related such that each contributes to the constitution of the other. Each field is characterized by its structures, including the time scales at which processes and events typically occur. On their trajectories through and across different fields, objects and people, marked by their own characteristic structures, find themselves interacting (being caught up) with objects and people, experiencing culture enacted with different structures. The relevant individuals in the school district offices may decide to have all students take a high-stakes standardized examination (in this case, the PSSA) and, on Monday morning, fax the instructions for Tuesday through Thursday testing. The fax considerably mediates events in the school and in the members' lives both on a short- and long-term level. In the short term, the exam mediates science teaching and learning as it disrupts classes for an entire week. In the long term, the outcome of testing has effects not only on students (whose college aspirations depend on these tests) but also for the school, whose very existence is at stake because the district office requires a certain minimum student achievement for the school to continue with its current teachers and administrators into the coming school year (Roth, Tobin and Ritchie, 2003). That is, boundary objects may create a mix of circumstances where individuals no longer feel in control of events but feel swept away, becoming reactive. The feeling of being swept away and being reactive is an effect of boundary objects, because participants in a particular field do not feel that they have control over these objects in the same way that they have control over objects that originate in their own field.

The arrival of boundary objects in some field is inherently associated with contradiction and conflict because of differences in structure between source and target fields (e.g., Hogle, 1995). Administrators (assistant principal, SLC coordinators) feel the need to deal with unfolding events in a more or less immediate manner, which leads them to experience themselves as reacting to all the problems emerging in the course of the day rather than attending to a previously established work plan. This is evident from the situations I encountered at City High. (From a research perspective, this makes any interviewing of coordinators and assistant principal during the school day a difficult affair, because any session could be terminated or interrupted by one of the frequently occurring reactive moments of indeterminate duration.)

Early Monday morning, the school had received a notice from the head office ("downtown") that PSSA testing would take place Tuesday through Thursday. The principals then communicated this information to the coordinators, who had to change their routines on the spot so that they could get all students to complete that portion of the standardized testing form that contains personal information items. To get the job done, the coordinator of SEM, Carambo, had asked three twelfth graders to assist him by taking the forms to the classrooms and then picking them up again once they were completed. Classes were interrupted and planned activities were rescheduled. That is, an artifact that had transited from one field into another brought about a substantial change in how the time available in the day was used. All parts of the school were affected including principals, coordinators, and students.

Carambo felt constrained in his actions, having to require all students in his SLC to complete the coversheet while attending to the normal demands of his job, which included the needs of different people as they arose in a normal day in the SLC. He described the effect of the memo from the district office as "PSSA nonsense" (see below). While he was personally attending to this task, he also had to deal with students being excluded from the classrooms by teachers and with teachers who had particular concerns and wanted to see him. As a result, he was prevented from doing what he had planned to do.

The district fax had effects on other objects, including the bell schedule (e.g., figure 4.1). The bell schedule is a boundary object for the explicit purpose of bringing different fields into or out of alignment by regulating their temporal unfolding. Students and teachers move from classroom to classroom, preparation and lunch periods, and begin and end their school day based on it. But between SLCs, lunch periods, for example, are scheduled differently (figure 4.1), which decreases the number of students in the hallways and cafeteria and therefore increases the levels of control over students that can be enacted. When the fax from the school district arrived, it mediated teaching and learning on this Monday and brought about the creation of a new bell schedule for subsequent days that was distributed to all members of the school community. That is, while the content of the fax had served the needs of district personnel to make the different schools and students comparable by aligning their schedules, thereby preventing information about the standardized examination from moving between different schools and students, it brought about considerable interruptions within the fields that it had targeted.

The school administrators realize that such changes as those brought about by the fax upset "a lot of people" (assistant principal) because, for example, teachers are asked to create lesson plans and they do, but then they are prevented from teaching according to their plans. Or teachers might have planned for and students have gotten ready for a unit test, which they can no longer administer when they are asked to attend a school-wide assembly without prior notice. The

	ECO					SEM			
	From	To	Min			From	To	Min	
	8 05	11 05	180	test		8 05	11 05	180	test
	11 05	11 18	13	adv		11 05	11 18	13	adv
	11 21	12 18	54			11 21	12 18	54	
	12 18	12 45	27	lunch		12 18	12 45	27	prep
	12 48	1 42	54			12 48	1 42	54	
	1 45	2 12	27	prep		1 45	2 12	27	lunch
	2 15	3 09	54			2 15	3 09	54	

E&L	11 21	11:48
SEM	12 18	12:45
CEO	12 48	1 15
ECO	1 45	2 12
ITA	2 42	3 09

PREP

Figure 4.1. Reworked bell schedule for two of the academies for the days on which the high-stakes examination were taken. Parsed out for the benefit of administrators are the preparation times for different academies.

assistant principal realized that different constituents in the school are frustrated when they are asked to change the schedule without adequate advance notice, and she also felt frustrated by the necessity to align the teachers' and her timelines with those of the school district.

> So it is an interesting dilemma because everything is reactive, because you don't have an opportunity to sit down and think. Well let's sit down and let's do that and do that. Everything comes at you. It is like being assaulted every morning. And there are very bad systems in place. (Assistant principal)

Carambo may be seen as reactive because he deals with the issues as they arise in real time. Having students complete the coversheet of the high-stakes exam is an instance of the timescale of the institution, shaping events in the classrooms, interrupting the flow of activity, and breaching the plans of science teachers. There is a contradiction in that an order that "comes down the pike" acts counter to the very goals of such tests—to hold teachers accountable for delivering high-quality education.

In this section, I show how objects that originate in one field become resources for the enactment of organization in another. They constitute a central aspect of organizational phenomena and directly address the question of how meso and macro level phenomena ever arise from micro level interactions. However, researchers should not assume that such boundary objects *cause* the events in a particular field. It remains a task of the analyst to show whether the enactment reproduces and reifies the organization or whether it destabilizes and thereby changes it. Thus, for example, there are circumstances where administrators allow some teachers and classes to not follow the request (boundary object) for all members of the school to gather in the assembly hall.

> I think a teacher should have the right to say, "Well, my class can't go." I brought that up at the leadership meeting, because we did [have such situations] last year. Several teachers came to me complaining because this [call for an assembly] came up very quickly, and they said they've got a test, and I said, "Well, I will go down and say you are not coming." And I'd say this is how you get teachers to cooperate more, because you are respecting what they are doing in the classroom. (Assistant principal)

Rather than assuming that instructions, orders, and rules produced at one (meso, macro) level of a school organization *cause* the behavior of its members elsewhere, researchers need to consider how boundary objects are taken up as resources in a field and mediate actions. The fit between action on the one hand and instructions, orders, and rules, on the other hand, can only be established, even for the most accomplished practitioners, after the fact (Suchman et al., 1987).

Crossing Times

Social structure is both a resource for and a result of social interaction. Through their actions in real places and under real and quite specific conditions of action, social actors structure schools as organizations. That is, schools are not fixed structures that exist outside, deterministically conditioning human actions and interactions. Rather, every day and in every instance, schools are reproduced, as they have historically evolved, and produced in new ways. Their inhabitants (administrators, teachers, students, support staff) continuously make and remake the organization what it is, all the while being oriented to what it historically has been. What needs explanation is this stability of organization in the face of the obvious human capacity to act this or that way.

Schools, like all social organizations, are unlike mechanical and simple physical systems whose behavior can be predicted fairly accurately. Unless there is some form of breakdown, a mechanized assembly line generates a continuous stream of products, one part of a machine interacting with another part in predetermined and infallible ways. Social systems, however, are never quite the same because they emerge from the dialectic of structure and agency, whereby structure itself embodies the dialectic of resources and schema. Dialectic means that there is an inherent, structural contradiction that leads to actions in a nondeterministic way. This is both a constraint, in that social actions are never quite the same and therefore are always associated with uncertainty, and an opportunity in that every social action can also contribute to changing the system. Most importantly, because of its nondeterministic nature, social action is flexible and can accommodate unforeseen circumstances. However, even the flexibility allows the reproduction and therefore reification of institutional structures as the following episode shows.

Thus far, I have shown that schools as organizations are made and remade when people use boundary objects, entities produced in another field, as resources for action in the field they currently inhabit. In the subsequent sections, I show how a school is produced and reproduced as an organization in face-to-face interactions, which constitute times when the trajectories of institutional actors from different fields align and cross for at least a short time. The episodes shed light not only on the work of members that make school the organizational setting what it is, but also on the work done by one member in particular,

Carambo, the coordinator of the SEM SLC. All three episodes show how organizational effects mediate science education in one form or another.

Making Time for the "PSSA Nonsense"

Events such as those requested in the fax from the school district office do not just happen; they are not the outcome of a causal chain of actions that characterize mechanical engines. Rather, such events have to be brought about through embodied human action. Furthermore, their very occurrence needs to be made possible. However, making possible the taking of tests or completion of forms requested by someone in the district office also produces the organization. When the requested event takes place, one can say that subordinates have followed an order or instruction. In this episode, I analyze how an event—the completion of the biographical section on a standardized test—was enabled during the interaction between Carambo and Solento, one of the physics/chemistry teachers in the SLC.

The episode was recorded on the Monday when I visited the school. As a result of this brief interaction, both Carambo and Solento made time for a task that became a twenty-minute disruption to a tenth grade chemistry course. That is, in this situation, they were not merely subject to time, but also produced time as a resource required in and for subsequent action. This time, however, would constrain Solento, who had planned to take students to the laboratory, and was preparing students for the lab work at the moment when Carambo arrived in her classroom,

```
 1   Carambo:   You have eleventh or tenth graders?
                (0.41)
 2   Solento:   I have tenth graders⌈
 3   Carambo:                       ⌊I am going to interrupt your class a
                little bit because they gave me all the P-S-S-A, P-S-A-T
 4   Solento:   ⌊Uh: um ⌋                                      ⌊Uh um⌋
 5   Carambo:   non sense to do.
 6   Solento:     ⌊Uh um⌋
 7                (0.40)
 8   Carambo:   and I've got to come in and take about
 9                (0.33)
10                Ten minutes so I'll . . .
11   Solento:   See, okay, so I, okay, after ten, fifteen minutes will be in the lab,
                so you're gonna have to come by=
12   Carambo:   =You be in the lab?
13   Solento:   Yea, so maybe you can, yea maybe one thirty five
```

14   Carambo: MAy be WHy I don't come ↓RIght away at
                     the beginning so yea so I come to your class first ⌈   ⌉
15   Solento: ⌊One thirty yea uh um yea   ⌋   ⌊yea o kay⌋
16   (0.25)
17   Carambo: ↓Aw'right. I'm gonna be right back.
18   Solento: °Yea, okay.°

In this episode Carambo and Solento, as social actors, make the organization what it is. They both use organization as a resource for their (verbal) actions and produce organization at the same time. Thus, Solento already acceded when Carambo, through his presence, indicated a wish of wanting to talk to her. Solento acceded and thereby made time for the interaction to occur, rather than saying, for example, that she was busy attending to the preparation of the lab. Making time to allow interactions to occur makes organizations what they are, even if these interactions are not planned but are made up and negotiated on the fly.

Carambo said that he was going to interrupt the class, but then placed the locus of control elsewhere, by indicating that an indeterminate "they" had given him "all the PSSA, PSSA nonsense to do" (lines 3, 5). Not only did "they" give him this nonsense to do, they are also powerful because they make him "come in" (line 8). "They," whoever it is, become powerful agents because they are able to make him, Carambo, do PSSA nonsense, all the while he qualified as nonsense the things that he was made to do. "They" not only have power over Carambo, but also over Solento, who is not merely asked to give Carambo some time to do the PSSA nonsense, but is told that Carambo would be coming in. How actors describe the organization is also how they orient to it; the two are sides of the same coin.

In this episode, the two participants not only made the time required for the completion of the forms but also reproduce institutional relations of administrator and subordinate. Carambo vied for control of the situation by saying what he would do instead of asking if he could interrupt the class for a period of ten minutes to get the "PSSA nonsense" done. In addition, he overlapped Solento's turn at talk, thereby vying for control over who was talking at the moment. Initially, he indicated that he was going to interrupt the class, beginning his utterance while Solento was still in the process of articulating the specific class she had (line 2-3). In this instance it looks as if Solento was at the end of her turn anyway, so that abandoning it was easy enough. However, later on, Carambo began to speak prior to Solento's completion of an idea unit by a process called latching, whereby he left no pause (lines 11-12). Similarly, in the course of Solento's next turn (line 13), Carambo began to talk (line 14). Solento did not show signs of abandoning her turn; at the same time, Carambo not only spoke much louder than Solento but also much louder than he had spoken before and after that. He not only vied for the turn by overlapping Solento, but as she did not stop talking, he increased the loudness, which only faded away when she had stopped talking (line 14). Once the situation was navigated and Solento no longer vied for the turn, the volume of Carambo's voice decreased toward the end of the episode (line 17).

Initially Solento indicated that she would be in the lab after the first ten or fifteen minutes of the lesson. Carambo therefore *had* to come by at that place rather than where they just met. Carambo would therefore not be entirely free to do as he wanted, but was constrained to come where she would be with her students. More so, Solento asked Carambo to come by the lab at a specific moment, 1:35 (line 13). But Carambo proposed a different course of action, according to which he would come to her class first—he elaborated on what this might mean in his next turn, "I'm gonna be right back" (line 17). As Carambo suggested that he was coming "right away," Solento began, overlapping with him, to revise the time to 1:30 and then acceded to his suggestion to administer the task "first" and "right away." Solento not only acceded by using the affirmative "yea okay" (lines 15, 18) but also by the decreasing loudness of her voice, which faded away in the last turn (line 18).

Stepping back, we can see this episode as a moment where time was being made for administering a task to students, filling out the biographical section of a standardized examination to be held on the next day. Solento was unaware of this new task, and Carambo was constrained in his action, knowing that this had to be done before the end of the school day. The analysis shows that Carambo made moves to impose the interruption according to his schedule rather than asking Solento when it would be most appropriate and opportune given that her class was in the lab. Solento ended up acceding, so that in this situation Carambo's institutional power as coordinator of the SLC was reproduced.

In this situation, not only Carambo's institutional power was reproduced but also that of the central administration, which controlled the events not only of the day but of the entire school week. The power is not inherent in "school administration," but it is inherent in the system in the sense that Carambo's and Solento's actions reproduced a structure whereby the instruction to have the standardized examination this very week actually got all social actors in the school to align themselves to this instruction. This was not always the case as my example of teachers and classes excused from a general assembly has shown.

Although institutional power was a resource for the conduct of social interaction, it did not *determine* how interactions unfold. Rather, there is evidence that Solento made moves to structure the interruption according to her own, here unarticulated, temporal commitments and structures. Thus, she actively worked for imposing her orientation to time and space, though in the end acceding to Carambo's schedule.

The outcome of the interaction had effects on science education in the sense that it directly constrained the laboratory activity, which Solento had planned (as per required lesson plan) and prepared. The completion of the forms taking place at the beginning of the science lesson, rather than when Solento had originally suggested, provided the opportunity for the process to take longer than the ten minutes that Carambo had forecasted. Thus, when all students were done, it had taken twenty minutes of the lesson, time that students could have spent learning science.

Setting Up a Child Study Problem (CSP)

Institutions are made not only in the recurring unfolding of nearly identical daily activities, but also in allowing special (series) of actions to occur when social actors recognize or establish a need for them. Because they are nonroutine, such (series of) actions have to be set up and organized (planned), requiring the mobilization and coordination of institutional resources including relevant personnel. Furthermore, the bringing about is itself an aspect of the organization, especially because it is not a regular aspect. Thus, we know that meetings such as getting a student into a special education service constitute dynamic events in which students come to be constructed as having some problem (learning disability, social problem) and plans of action are established (Mehan, 1993). The institution may or may not have established protocols for conducting such meetings, which are resources for action rather than causal determinants of actual events. However, establishing the need for such a meeting is itself an aspect of institutional work that social actors do, and for which they make time, often in an ad-hoc fashion. Brief and innocuous interactions in a hallway or classroom may contribute to the constitution of institutional processes that unfold with their own temporal dynamic and lead to resources (e.g., plans of action) that constrain the actions of others. Institutions therefore are not just made in important meetings or processes with a preestablished order for unfolding, they are also made in the short and insignificant moments when constituent actors decide to get some other process started. I exemplify these interactional aspects in the following episode, which led to a student becoming subject to a process called the "child study problem" or CSP.

According to Carambo, the first tier of a CSP involves the student's parents, all his or her teachers, school nurse, counselor, special education teacher, and

SLC coordinator. The purpose of a forty-five minute CSP meeting is to arrive at a plan of action, a coordinated series of steps designed to ultimately lead to improved grades or social behavior. If these interventions turn out to be insufficient for addressing the problem, a decision may be taken to move to second-tier processes, which normally include additional professional help.

In the following episode, Carambo and Solento make, almost in passing, a substantial decision that brings about a variety of events centering on a student failing physics and chemistry, and which involves different people, normally spread across space, who may seldom see one another face to face. In the process, they produce and reproduce the institution, and reify the CSP as a viable process for dealing with particular issues. More so, even this brief encounter concerned with a student and his problems, is a moment for constituting institutional relations between the SLC coordinator, on the one hand, and one of his teachers, on the other. In the process, they produce and reproduce school, organization, power relations, and their own identities.

```
 1   Solento:    And about Carley?
 2               (0.52)
 3               °↓Okay.
 4               (0.23)
 5   Carambo:    WHo?
 6   Solento:    Carley.
 7               (0.21)
 8   Carambo:    °↑Eche°vari a?  ⎤
 9   Solento:              ⎣Uh ↓m.
10               (0.31)
11   Carambo:    °U:m:°
12               (0.90)
13   Solento:    Doesn't want to do anything. He is failing
14               (0.29)
15               like both class es, chemistry ↑anD physics.
16   Carambo:               ⎣Good, so::?  ⎦
17               Want to do a C-S-P?
18               (0.55)
19   Solento:    Yes.
```

Immediately prior to this episode, Solento and Carambo had talked about another student, who had caused some problem in Solento's class because he wanted to sit in the back of the class and talk rather than in the front where she had asked him to sit. Only a brief pause after Carambo had articulated an action he was to take with respect to that student, Solento uttered "And about Carley?" which, because of the rising tone at the end, can be heard as a question, "What do we do about Carley?" There was a 1.15 second delay before Carambo spoke, bridged by Solento's utterance "Okay" (line 3). This utterance was not designed to take the next turn. Carambo took his turn with an especially loud "Who?," which can be heard as an indication that he was unclear both about content of the question, who the particular student is, and context, as if he was returning to a conversation that he had not attended to. The two elements together constituted a request for providing more specific information. In response, Solento simply re-iterated the first name of the student, with emphasis (line 6). In this, she indi-cated that Carambo ought to know the particular situation referred to, and per-haps that she took the long delay as an indication that Carambo had not heard the name. Carambo then uttered a last name, which Solento confirmed even as Carambo finished pronouncing it (lines 8, 9). There is then a short and a long pause, surrounding Carambo's "Um," which indicates that he does not want to take the turn but is perhaps waiting for the problem to be further articulated by Solento, who did so by elaborating that Carley was not working and that he was failing both of the courses that she was teaching him (lines 13, 15). There is then a pause, during which Carambo could have started a turn but did not. A pause can serve as a request for further elaboration and provide the space for it. Once Solento's unfolding articulation was beginning to sketch the problem, Carambo utters what can be heard as a request for her solution, "Good, so?," and then he specifies whether she wants to have a CSP (line 17). After a (hearable) pause, Solento affirms.

In this situation, the two not only brought about a CSP for Carley but also enacted power in an asymmetric way. Although Solento identified the particular double problem at hand, the student did not want to work *and* was failing, it was Carambo who proposed the CSP. Equivalently, Solento might have requested a CSP. If she had asked for a CSP, she would in fact have attempted to make a de-cision for which Carambo was responsible. Solento did not. His authority there-fore had been reproduced and went unchallenged.

Despite its apparent brevity (less than twelve seconds), this unplanned meet-ing would have long-term consequences in the school and to the student. It is not just that these brief interactions bring about the alignment of temporal processes of different fields and of different scale, they also perturb other fields, such as the programs that are made available to students, and therefore the development of students. In this brief interaction, a decision was made to bring together the student's teachers, nurse, counselor, parents, and special education teacher to talk about a plan of action. Such a meeting results in a different trajectory for the student for the next weeks, months, and even years, perhaps until he completes high school. Yet more work needs to be done before this meeting can come about. Time and space are resources that need to be made available, necessitating that the schedules of the quite diverse membership are synchronized.

Whether the CSP would lead to an improvement in Carley's grades cannot be known. Nor can it be known how the CSP would mediate other aspects of Carley's life, reproducing failure or producing a very different trajectory altogether. Concrete analysis of the specific case is required to understand how particular structures created by actions become opportunities and constraints for subsequent actions. My research at CHS showed that even students apparently doomed to failure can participate in creating resources that ultimately lead them to successful high school and college careers.

Producing the Caring Coordinator

As in all schools, there are many opportunities for contradictions and conflict to mediate particular fields, which prevent events from unfolding in the ways they are normally reproduced with little change. How conflict is dealt with contributes to making the organization what it is. In SEM many students and teachers recognize Carambo as a person who deals particularly well with students, especially in conflict situations. Carambo's orientation toward action is one that focuses on students and their needs.

Carambo's "students first" orientation would not have much purchase unless it was the outcome of interactions. Here, I analyze a brief interaction with Bobby, one of the students mentioned in the introduction to the chapter. Together with another student, he had been sent out of his classroom because of behavior perceived by the teacher as inappropriate. Due to a contradiction in the way students are assigned to courses and classes, both students felt familiar with the content and were bored as the teacher attended to the needs of students insufficiently prepared because they lacked a prerequisite course. (A detailed analysis of these contradictions and the mediating effects to science learning is provided elsewhere [Roth, Tobin and Ritchie, 2003].) Whereas the other student attended to his homework for another course, Bobby, mediated by the onset of a flu infection, went to sleep.

| 1 | Carambo: | * Bobby what's wrong with you man, you don't feel good?=You don't look good, first of all. |  |
| 2 | | (0.46) | |

3		Did you? WAIT, WAit, *
4		(0.83)
5		What's the matter?
6		(0.37)
7	Bobby:	Oh I got a little headache is . [. .
8	Carambo:	[You] look funny.
9		(1.03)
10	Bobby:	I've got a headache.
11	Carambo:	You're giving him the blues in there a little
		bit in that class, isn't you? *
12		(1.65)
13		So when we finish me and you gonna go
		back and chat with him a little bit
14		(0.33)
15	Bobby:	Uh um.
16	Carambo:	You can't sleep in there but you can sleep in
		the * back corner over there.

Here, the student Bobby showed up in the science classroom that Carambo currently uses as his headquarters. Not only did Carambo know that there was a problem in general, but from a brief exchange with another student immediately prior to this episode, he was aware that Bobby had been sent from his science class because he had been sleeping. Carambo did not ask for an account of the events but inquired about Bobby's health (line 1). Bobby did not answer but turned around and began to walk away toward the back of the room. Carambo not only asked him to wait, implying that the conversation was not concluded, but also reached out, taking the student by his arm, and thereby stopping him from proceeding (line 3; the asterisk coordinates text and offprint). But Bobby did not explain. There was a long pause, which, "predictably" (Boden, 1994), led to a reformulation or rather elaboration, "What is the matter?" Another brief pause followed before Bobby explained that he had a headache, an utterance that Carambo overlapped at the end by articulating a perception, "You look funny" (line 8). Bobby reiterated having a headache, and Carambo, now physically close to the student, asked in a conciliatory tone whether he was "giving [the teacher] the blues" (line 11). As there was no response for a considerable time, Carambo proposed a course of action of going back to "chat with him [science teacher]" (line 13). Bobby acceded, and Carambo elaborated that the student could not sleep in the science class ("in there") but offered him the chance to sleep in the corner of the room where they were. As he uttered his offer, Carambo stepped even closer to Bobby, pulled on his side to encourage him to turn to the back of the classroom and gesticulated "over there" (line 16).

An important outcome of this interaction was the establishment of respect and a caring attitude. Although Bobby had been kicked out of his science class, he did not use the occasion to let off steam or to complain. When Bobby did not articulate his concerns, Carambo persisted in finding out matters concerning the student and proposed a course of action that itself foreshadowed mediation rather than punishment. Even for the time to be spent outside the classroom, Carambo offered Bobby the opportunity to sleep rather than asking him to engage in some irrelevant task for punitive purposes. Carambo emerges from such interactions as an administrator who cares, and students respond, as Bobby, by avoiding conflict.

The decision to have a chat with the science teacher is in fact a resource for dealing with the contradiction that had arisen. The chat led to a cogenerative dialogue, a form of interaction that provides all participants opportunities to contribute to a better understanding of the events in which they are caught up, and to resolve issues. That is, the interaction not only produced Carambo as a caring coordinator but also created the possibility for Bobby to return to the science class and to continue learning rather than to lose face and to regain social capital through actions that somehow undermine the science teacher and constrain the science lessons.

Although episodes such as this lead to the construction of Carambo as a caring coordinator, and therefore the partial construction of the organization as caring, an asymmetrical relation was clearly reproduced toward the end of the episode, as Carambo articulated what Bobby could and could not do. Furthermore, when Bobby clearly wanted to stay out of an explanation—he turned to follow his classmate—Carambo stopped him by firmly but not violently preventing Bobby from moving away. At the same time, the action enacted closeness and caring. Carambo physically stopped Bobby, then, when the student did not answer by a direct question about his state, Carambo inquired, "What's the matter?" (line 5).

Production and Reproduction of School Organization

Science educators usually deal with knowing and learning in classrooms as if one could usefully separate them out as a field (unit of analysis) that is not influenced by events and structures elsewhere in the larger organization. What and how students learn and what and how teachers teach are then problems of individuals or, at best, of the classroom as a collective. In the present study, I show that fields interact and the events and entities in and from one field shape what happens in other fields, where they come to structure actions, being deployed as resources and schema. More so, I show how some organizational features of school structure emerge from the interactions of people in particular fields, the interfaces between fields, or by attention to objects that move across boundaries and contribute to shaping events in fields other than those that produced them. The chapter therefore shows how micro, meso, and macro level aspects of urban schools are different aspects of the same phenomenon; they are outcomes of collective life, continuously instantiated in human interaction.

Theorists have long thought that organizations are most efficient and productive when they are split into distinct physically separate and compartmentalized fields distinguished in terms of power, hierarchical leadership, and distinct status of individual members (Lindbeck and Snower, 2000). All of these features are consistent with a fixed and static conception of organization. However, I show that without the movement of objects and people and without the fluidity of face-to-face interactions, schools would not be what they are. The very fluidity embodied in the dialectic of structure and agency is a resource not only for reproducing schools as the rigid organizations that they are but also for producing them in new ways, more congenial for enacting their raison d'être. That is, organizations are not mechanical structures that are irrevocably fixed in how they operate and what they produce—although they frequently exhibit predictable collective actions, this predictability is itself an outcome achieved in indeterminate person-to-person interactions. Flexibility is a positive phenomenon, because it gives us hope for changing situations, which we recognize as not working, and in fact, is consistent with a historical perspective on organizations, which evolve in time. Inflexible school systems would not be able to accommodate changes in external conditions or in response to internal contradictions.

Neighboring (linked) fields are often treated in educational studies as having little effect on one another (Lemke, 2000). For example, studies of learning science regularly look at processes not only independent of the school as a whole but also independent of processes that make a classroom what it is. However, in many instances, the coordination across fields is an organizational reality fraught with difficulty and contradictions. In this study, an instruction created in and disseminated by the head office of the school system becomes a resource that substantially interferes with the events in the school at multiple levels: the bell schedule had to be changed, teachers could not teach the lessons they had prepared, and students missed out on their science lessons. This is a very common phenomenon in human social activity and many aspects of social life can mediate the integration of social activities across different settings and timescales. Visual artifacts, architecture of rooms, layout of seats with respect to focal artifacts mediate events in the settings on smaller timescales such as the structure of the turn-taking in talk (Roth, McGinn, Woszczyna, and Boutonné, 1999). They constitute material constraints or opportunities for interaction specifically and for making the school (as an organization) what it is in the daily praxis of its inhabitants.

Boundary objects—e.g., instructions from the school district office, or bell schedules for students—may give social actors the sense that they are reactive. The instruction generated in the school district office to administer a standardized examination had an influence on Carambo's day so that he had to stay to do after school what he, mediated by the "PSSA nonsense," was prevented from doing during the school day. How they deal with boundary objects mediates their careers, whether they are students, teachers, or administrators. Feelings of having to be reactive have repercussions for administrators; these feelings themselves are resources for action. If they do not accomplish what they had planned because of a need to attend to the contradictions and conflicts arising from aligning and accommodating objects stemming from and differing in temporality, often administrators will do their work when others are not in the building to disrupt them (e.g., before teachers and students arrive in the mornings, after they have left in the evening or on the weekend). Principals and some coordinators often find one another back at the school on weekends to "catch up" with their work. In fact, extending normal working hours is interpreted as an indication of "commitment," which becomes an interpretive resource when individuals are considered for new or vacant positions. This is particularly the case since this commitment is recognized as being at the cost of time to the family life, another field with its own temporality and time demands, rhythms, cycles, pacing, and events. That is, commitment, expressed in the willingness to give "extra time" (i.e., volunteering to tutor students during lunch and after school), translates into changes in another field, personal career trajectories (cf. Brett and Stroh, 2003). In fact, there were indications that administrators not only viewed science teachers as unprofessional (because they did not give extra time to tutor students after school, valuing family time more than an additional commitment in and to school) but also wanted to get rid of them—which would have tremendous influence on science teaching and its continuity in the school.

The bell schedule is another boundary object that is associated with contradictions and conflict. Students have little control over its construction, but its effects mediate practices that interfere with science teaching and learning. Thus, the bell schedule is a resource that not only contributes to the reproduction of schooling structure, but also to practices that disrupt and question it. For example, students who "come late" to science class show that they are masters (in control) of their time rather than being reactive. These students question the institutional structures built on a conception of time as something that not only unfolds linearly but also has become a commodity ("time is money") which can be gained and lost. From this forced alignment, a number of contradictions and conflicts arise, such as when individuals more attuned to social time become aligned with a linear, machine-like, clock-time-driven schedule. Here, institutional power that attempts to entrain students into the bell schedule is explicitly undermined. School as an organization willing and able to educate the students is questioned in its very raison d'être. At the same time, because the school holds students accountable through testing, failed tests, and low grades, by means of a chain of events and processes, ultimately leads to a reproduction of the underclass status of the students, their families, and cultures. Hence, in science classrooms and school in general, contradictions are experienced by African American youth, who place high value on communalism, social connectedness, and social time (Allen and Boykin, 1992). The greater value placed on such characteristics supports practices that are often in conflict with those associated with an adherence to clock time.

Schools as organizations stand and fall with the face-to-face encounters and meetings that produce and reproduce them. During such crossing times, different stakeholders and peers can get themselves "onto the same page," by aligning their visions and actions. Informal and formal, brief and extended, and unplanned and planned meetings are at the heart of negotiating and coordinating the different fields and temporal zones. Agendas, actors, objects, times, and places vary, but meetings are the proper organizational activity for management, locating and legitimating both individual and institutional roles. However, the alignment of different temporal zones and material resources, necessary to bring about a common meeting, is itself fraught with difficulties. It is difficult to schedule meetings, but meetings are the place where conflicts arising from temporal misalignment can be dealt with. I am certain that organizational studies will provide us with better understandings about the successes and failures of schools and schooling, and in particular the poor job society has done to generate more opportunities for the students that currently populate urban schools.

There Is Hope

In this chapter, I develop a view of schools as organizations that emerge from the movement of boundary objects and face-to-face interactions. Some readers may now ask: "How is this way of looking at school organization helpful? How are the understandings you develop useful to teachers and administrators? and Can teachers do anything other than reify and reproduce structures such as the CSP, since they represent school policies?" I begin by answering the last question as my response will also answer the previous ones. Teachers can do more than reify and reproduce structures; in fact, they continuously produce structure since every action, however similar to the previous one, is inherently different. All cultural phenomena undergo continuous, though often imperceptible change, which is brought about *because*, in acting, we also produce rather than merely reproduce structure. This perspective gives us, as urban educators, the hope we need in our work so that rather than taking schools and school districts as given conditions that we have to accept, we have the possibility and capacity of changing these conditions. This allows administrators, teachers, and students to understand that our actions, no matter how small, have effects: they produce and reproduce the organizational features of schools, including such programs as the CSP. We come to understand that things could be otherwise; and we come to understand that actions, such as those involved in "doing a CSP," create resources and constraints which can have lasting impact on the person who is the target of the actions.

The perspective on schools taken in this chapter teaches us about the fundamentally dialectical relation between schools as organizations and the individuals (teachers, students, and administrators) that constitute them. My message is not only that we *re/produce* our schools but also that *we* re/produce our schools. Thus, each individual can contribute to changing both the schools and school districts, which currently contribute in significant ways to the reproduction of inequality. However, we cannot do it on our own: collective action and interaction change the practices rather than individuals. Even though an individual administrator may institute a new policy, its practical realization depends on the collective. Policies are only structures; they, too, are kept alive through reproduction and can be changed through our actions.

Ultimately, the perspective makes us administrators, teachers, and students aware of our individual and collective responsibilities. Our actions continuously produce and reproduce social structure and, therefore, we can always do otherwise. We therefore have the responsibility of choice. Do we want to contribute to the reproduction of poverty, inequality, and injustice that characterizes our society or do we want to contribute to producing a different world?

Editors' Perspectives

This chapter examines science education in an urban high school through micro, meso, and macro lenses. An important issue for researchers is to address macroscopic social issues, such as those that encompass more than one field. Social life is clearly more complex when individuals have to deal with different structures in the fields they populate, especially when those fields intersect or are nested within one another. The examples dealt with in this chapter are classic instances of nested and intersecting fields and the contradictions that arise when resources from one field are accessed and appropriated in another. It is apparent that adequate accounts of social life necessitate explorations of the manner in which interconnected fields structure social life and thereby afford the agency of participants.

Roth contributes to the theoretical underpinnings of the study by regarding a school as a field consisting of an organization that contains numerous fields. In this way he is able to show that contradictions can be created by the different structures extant in a field. His study of temporal issues underlines their salience as factors that have not been studied as extensively as they might be. Additional studies of time, as a component of structure, are considered a priority for research in urban science education.

Uses of cogenerative dialogues have proven useful in identifying contradictions experienced by participants in a field. In instances such as this one in which many fields are nested and intersect, cogenerative dialogues should include participants from each of the relevant fields so that any collective decisions that are cogenerated can include participants with the necessary capital to initiate practices that will allow those decisions to be enacted to produce successful outcomes. For example, decisions about the roster are vital to the learning of students, teaching, and the efficient use of human and material resources in a school. Hence cogenerative dialogues should occur among stakeholders to identify contradictions and resolve them.

Chapter 5

Playin on the Streets—Solidarity in the Classroom: Weak Cultural Boundaries and the Implications for Urban Science Education

Rowhea Elmesky

It was another day at work and three of our high school student researchers were engaged in a science-related activity as part of their employment responsibilities. We were doing the well known "Egg Drop. " I provided them with various materials, including cotton, fabric, balloons, Styrofoam, cardboard and to design a protective module in which an egg could be placed and dropped from a high location. An unbroken egg represented the immediate goal, yet the deeper intent was to help the group to learn about forces, velocity, and acceleration by measuring the distance the egg was dropped and then recording the time period for the drop. More specifically, the activity was designed to encourage discussions around Newton's Laws of Motion. On the first morning that the project began, the youth were working on their designs when Ivory decided to narrate a humorous story. She directed the conversation to Akram, a young African American graduate student who was hired to support the research project throughout the summer (and interned at City High the following fall). His primary roles included supervising the students while they were not working with me, videotaping learning sessions, and guiding the students' development of teacher resources, for example, their movie on "Sound in the City." Suddenly *playin* erupted in the midst of the science activity. Akram was holding the video camera, and Ivory began to narrate an incident she had witnessed as a car drove by her.

Ivory	The lady was on a truck.
Akram	(*Using an exaggerated loud voice*) Is it because I'm black?
Ivory	She ain't have no [teeth
Akram	[Was she black?
Ivory	Yeah. (*Starts to rock her upper body backward and forward in imitation of the movements of the woman she is describing.*)
Akram	(*Using an exaggerated loud voice, again.*) Is it because she's black?
Ivory	She was like. She was like "Yeah!" (*Stops rocking*) I think the song—whatever song it was—she was like: (*Ivory starts again rocking back and forth in large sweeping movements in her chair and claps her hands twice as she reenacts the lady's movements as well as what she was saying.*) "Yeah come get me! Yeah, come get me!" (*Laughing*)
Akram	That's how your aunt be? (*Momentarily, it is quiet as Ivory's wide smile diminishes into pursed lips hiding a slight grin. Then another student researcher and Akram erupt into loud laughter in synchrony, and Ivory's grin continues to grow until she joins the laughter.*)

*

Like me, I have a quick tongue . . . I really do, an' for me to suppress it, is hard. But I have to do it often. An' like my quick tongue, it can be intelligent. It can be sarcastic. It can just be. (Natalie, interview, 2002)

*

As described by Natalie and demonstrated by Akram in the opening narrative, being able to verbally *play* (i.e., having a "quick tongue"), or embodying a talent for creatively and artistically utilizing language in manners that are "intelligent" or "sarcastic" to express oneself or to respond swiftly and adeptly to another's statements or narrative, is a practice enacted, often unconsciously, by many of the youth from City High as they interact socially and/or academically. This research shows that in addition to underlying goals of building and maintaining respect, *playin* is also utilized in building relationships, social networks, and "solidarity" (Collins, 1993) in terms of group membership. When engaged in classrooms and schools, or in the midst of a science activity as described in the opening narrative, these practices are often mistaken as forms of resistance or become perceived as conscious methods for disrespecting the teacher or other students, thereby evoking disciplinary responses.

Afro-American academic difficulties were typically explained in terms of the student's own inadequacies and problems. If black children do badly in school, we must discover what is the matter with them: they may have maladaptive reactions to adversity or inadequate socialization experiences, especially at home. (Boykin, 1986, 60)

Unfortunately, in urban classrooms, where most of the students come from social and cultural histories that differ from the teaching body, there is not only a dispositional disconnect between teachers and students but also a concerted effort of teachers and administrators to "make" these students fit the mold of a "model" student. This is particularly true for economically disadvantaged African American students (like Ivory and Natalie), since different realms of experience, including "African-rooted black cultural" and "minority" experiences (Boykin, 1986), help shape some of their dispositional resources and may place them at variance to mainstream culture that determines what is "school-appropriate." Rather than trying to understand student practices that seem to be contrary to school culture, there exists a rejection of nonmainstream ways of being. Moreover, while it seems obvious that the negotiation of these different ways of being is in constant breakdown in urban classrooms, evident in the excessive focus on classroom and school management and the demand that urban students work to overcome so-called "deficiencies," there has been little research that attempts to understand how macro issues, such as growing up as a poor minority in a society structured by Eurocentric values, shape teaching and learning in urban classrooms. Nor do we have research that investigates how differing teacher/student cultural and social histories contribute to the dynamics of classroom solidarity and student opportunities to exercise agency in accessing resources that will contribute to their success as science learners. Moreover, while the literature does speak strongly to the lack of resources in urban schools, in the form of both materials and qualified teachers (i.e., Haberman, 1991), what remains sparse is research that looks at the rich human resources that are embodied within urban youth in inner city schools and ways to re-envision how we think about science teaching and learning so that this capital can be utilized in the planned and enacted curriculum.

In this chapter, I present understandings of how dispositions to oral tradition, expressive individualism, verve, and movement (Boykin, 1986) can manifest through different forms of verbal *playin*. I argue a both/and perspective in which I choose to look at the world in terms of dialectics rather than dichotomies (K. Tobin, personal communication). Thus, these dispositions, which may be utilized by youth in social fields outside of school to accomplish a multitude of goals, can also be viewed as an untapped reservoir of resources for affording students' collective, as well as individual, agency as learners in the science class-room. Accordingly, this chapter strongly advocates that *playin* dispositions be reconceptualized as a resource for building positive emotional energy (EE) or "a feeling of confidence, elation, strength, enthusiasm, and initiative in taking ac-tion" (Collins, 1993) around being in science classrooms, and for contributing to the development of identities aligned with being able to *do* science and develop-ing scientific fluency (Tobin, chapter 2).

According to Boykin (1986), oral tradition as a cultural disposition is "a preference for oral modes of communication in which both speaking and listen-ing are treated as performances and in which oral virtuosity—the ability to use alliterative, metaphorically colorful, graphic forms of spoken language—is em-phasized and cultivated" (61). Whereas Boykin writes of oral tradition in a man-ner that sounds very conscious in nature, Natalie's expression, *"it can just be,"* indicates that there is an unconscious aspect of oral tradition which manifests in the form of verbal *playin* dispositions. Students like Natalie are often forced to make conscious much of what is unconscious so that they may successfully navi-gate classrooms, day after day, throughout their school career. There are real consequences for having a "quick tongue" in her high school, particularly since teachers often misunderstand the underlying playfulness of her verbal exchanges (causing her to have to point out: "It's not that serious!"). In contrast, Natalie pleasantly recalled one course in which she had a student teacher, Akram, whom she believed understood and embodied such dispositions to *playin.*

> Mr. Akram *know.* Mr. Akram used to just stand in the doorway just so I could bump him cuz he knew I was gonna do it whether he was in the doorway or not. Mr. Akram was so much fun. He was one of the best student teachers I had because he always found a way to make the class enjoyable.

Arguably, it is perhaps through a shared understanding of cultural practices such as *playin* dispositions that Natalie and Ivory are able to experience high levels of solidarity with a teacher/research assistant like Mr. Akram, which they couldn't have otherwise. In schools—particularly ones where the teachers and students have differing social and cultural histories—it is important for students like Natalie and Ivory to have interactions from which they can experience a sense of solidarity and build positive emotional energy within their classroom, with each other, and with the teacher.

Overview of the Research Approach and Conceptual Lenses

This chapter provides macro, meso, and micro findings emerging from a study on improving the teaching and learning of urban, economically disadvantaged youth. African American student researchers (Shakeem, Randy, Ivory, and May) from City High, a comprehensive neighborhood school, developed most of the ethnographic resources that inform this chapter. Informing this research are multiple data resources, including (a) raw video footage taken by student researchers in their Philadelphia neighborhoods, (b) edited video ethnographies produced by student researchers, (c) audiotaped conversations and interviews, (d) field notes and video records of time spent in different social spaces outside of school with the youth, (e) videorecordings of science classes at City High, and (f) videorecordings of science-centered research activities during intense summer research sessions.

The approach to understanding data differed from traditional ethnographic research, in two distinct ways. First, in accordance with my theoretical lenses, which value both coherence and contradiction (Sewell, 1992), my analytical approach stressed the importance of representing social reality in its complexity, so that both the patterns and contradictions were identified. Second, my approach emphasized the importance of engaging micro-level analyses which allowed much more detailed aspects of a field and interactions to be understood. I used video editing techniques to capture (and convert into digital form) vignettes that supported or contradicted meso-level observations of what appeared to be *playin* practices, according to student researcher definitions, in a variety of fields. Selected from video footage of science classrooms, the university research setting, neighborhood streets, students' homes, and other social spaces, I then re-examined the various clips on a micro level by varying the speed, from frame by frame up to twice the regular speed to identify thin patterns of coherence of *playin* as well as the contradictions (Sewell, 1999). I also watched closely for nonverbal cues (i.e., smiles, eye contact, body rhythm) associated with *playin* within the selected vignettes and searched through other data resources such as interviews to determine further patterns or contradictions.

> Sometimes it's different types of quick tongues. It's a sarcastic one, it's the one—the sarcastic one is more or less the common one. It's like whereas then they have the one where they have the intelligent quick tongue where they say OK. Well, someone will say something like, "Well, you got a McD, well that's just more rice for me!" . . . Or my nephew Ed, he'll say somethin' like "you gettin' on my nerves." [I'll respond] "I love you too Ed." It all depends on the surroundings and the conversation. (Natalie, interview, 2002)

In conducting the data analyses necessary for writing this chapter, I utilized student researcher constructs to identify and categorize examples of *playin* or the use of a "quick tongue" across different social fields. However, while student researchers directly and indirectly informed this research and while the artifacts and date resources are deep and rich in capturing the unfolding nature of events and the role of *playin* practices therein, I recognize that this chapter is a construction of my own narrative. In telling this narrative, I incorporate a blend of cultural sociology (e.g., Sewell, 1999) and the sociology of emotions (Collins, 1993) to make sense of the unconscious nature of *playin* practices engaged by some inner city African American students as they participate in a variety of social spaces including their science classes. I utilize culture as a construct that can be dialectically conceptualized as a system of symbols, the associated meanings, and practices (Sewell, 1999) and that is loosely bounded so that practices originating in one field may appear within another. Although practices are not always unconscious in nature, particularly when they break down (i.e., you are unsuccessful in mediating your goals in a particular field), this research connotes *playin* as part of an unconscious system of dispositions or habitus (Bourdieu, 1992) that is shaped by and also shapes the social spaces where culture gets enacted (fields).

Many students growing up in the inner city have lived experiences in which they have come to embody *playin* dispositions for various purposes. Furthermore, from being within fields which are structured in ways that encourage interactions marked by verbal *playin*, and due to the weak boundaries of the fields, the meanings associated with *playin* dispositions can transcend classroom walls and, depending upon the classroom structure and the actors involved, afford or truncate student agency (Sewell, 1999). When actors share dispositions such as *playin*, there exists a strong possibility that their interaction will be synchronous and produce positive emotional energy (Collins, 1993). Utilizing theory to form connections between dispositional enactment of culture, synchronous interactions, and scientific fluency, this chapter answers important questions of how *playin* practices afford student agency and can improve teaching and learning in urban classrooms.

Macro Structures That Afford *Playin* Practices

In this study, I situate meso and micro understandings of *playin* practices within larger, macro-level understandings of the types of societal structures that afford the development of such ways of being. More specifically, the following section addresses the ways in which poverty shapes the lived experiences of many inner-city youth and situates our understandings of the centrality of respect on inner-city streets (i.e., Anderson, 1999) within more central constructs of solidarity and social networking.

The Need for Solidarity and Social Networking

Inner-city neighborhoods are marked by a dramatic lack of material resources. Manifested in widespread poverty, high crime rates, and the proliferation of drugs, these aspects of structure shape the ways in which inner-city youth can access and appropriate resources in the dominant culture, and, as a result, patterns associated with an "underclass" have emerged, characterized by a lack of jobs and social isolation (Wilson, 1987). Nested within a history of decades of inequitable political decisions and policies, inner-city neighborhoods remain trapped in such cycles of intense poverty and racial isolation (Anyon, 1997). Living within conditions that suggest a lack of societal concern for the urban poor, this research suggests that youth have found their own resources for navigating their lifeworlds and exercising agency or direction over their lives. In fact, the isolated, segregated aspects of inner-city neighborhoods reinforce a structure that tightly binds communities together and supports the emergence of communalism or a "commitment to social connectedness which includes awareness that social bonds and responsibilities transcend individual privilege" (Boykin, 1986, 61). Moreover, there exists ideology and practices that promote the development of youth who exhibit high levels of solidarity and loyalty, and these constructs are often significantly intertwined with the importance of having status and being respected by others. Within the repertoire of practices associated with communalism, *playin* has emerged as an important component of relationships, an instrumental resource as social capital is built and maintained among many of the youth attending City High.

> From boys to men; from the end of the road.
> Yo, we boys to the end, never been out in the cold.
> You got me. I got you. Before our souls grow old.
> He shot you. He shot me. I was supposed to roll.
> I take a shot for my nigga. Give me two to the ribs. . . .
> Know what I want in my life?
> I want for my brother.
> Know what I want for my wife?
> I want for my mother.
> It ain't a question of what I wouldn't do for my squad.

Ask yourself if you really truly a squad.
I'm gonna ride with my niggas.
Die with my niggas.
Get high with my niggas.
Flip pies with my niggas.
Til my body get hard.
Soul touch the sky.
Til my number gets called an' God shuts my eyes.

For Shakeem, the rap lyrics of Beanie Sigel's "The Truth" are a powerful expression of what solidarity or group membership means for urban youth living in the nation's most economically disadvantaged neighborhoods. Shakeem has been through a variety of challenging experiences in his sixteen years of life, and he has been able to share some aspects of his lifeworld while working as a student researcher with our project for the past two and a half years. In his edited video ethnography, Shakeem incorporated music and text with video footage to share "who he is," by highlighting some of the dispositions he identified within himself through the analysis process. By watching Shakeem's ethnography and through working with him over an extended period of time, he has taught our research group about the different types of relationships that youth may form in inner-city neighborhoods, the various meanings associated with each, and the value that true friendship holds. As the above lyrics play in the background, Shakeem's commentary rolls slowly across the screen, "I have many associates. I have a few friends. But only one nigga. My nigga Cas." Thus through the powerful combination of words, text, and video, Shakeem vividly brings to life the concepts of solidarity, community, and friendship, captured in Boykin's (1986) reference to communalism.

The type of solidarity that holds a "squad" together on the streets is one that transcends the privileges of oneself and reaches the level where what you may want for your "nigga" is what you would want for your immediate family of loved ones. This orientation in inner-city neighborhoods to communalism and solidarity is linked in many ways to issues of safety. In comparison with its nearby suburbs, inner-city Philadelphia has a murder rate that is seven times higher. In fact, according to the Heritage Center for Data Analysis (2000), young African American teenage males tend to be the group overrepresented in these statistics such that in Philadelphia, the probability of an African American teenage male, like Shakeem, being murdered before reaching the age of forty-five is 5 percent, or about one out of twenty are murdered. In an article recently published in the *Philadelphia Inquirer* (12/13/03), the homicide rate has been noted to have a 15 percent increase over last year. Considering that Shakeem has already been shot at (and missed), it seems smart for youth to engage in practices that will build strong bonds of friendship, ensure their well-being, and possibly ward off harm. Thus, social networking is not only very important; it is often necessary.

The need for building social capital is not limited just to males, as females may experience challenges to their safety as well. For example, recently, May's family had to relocate within Philadelphia to escape harm that was being inflicted by her neighbors. During one of our intense research summer programs, May was found to be continually missing morning after morning. Although she would call to report that she would not be able to come in for work, it was unclear to the university researchers involved as to why she was unable to maintain a regular schedule. During dialogues that occurred between her and me during the following fall, she was able to shed some light on the matter. She summed it up in a sentence. "They was crazy around my old way," she explained, and May's family had few social networks to protect the family from harm. For months, May's younger sister was being "roughed up" by a neighborhood girl, until she finally fought back. This provoked the female, who had been instigating a confrontation, (along with a group for support) to come after May's sister for retribution. May became involved since, as she pointed out, "I couldn't just stand there an' watch all them girls jump on my sister. That ain't right." Another night, May was in the emergency room with her older sister, five months pregnant, who had been kicked in the stomach during a physical altercation. In these types of situations, May explained that having someone, or a group, for support is important. She stated, "You never know when you need them. It's good to be close to people." Moreover, as Shakeem expressed, having "a good nigga at your side" or someone to be there through the thick and thin is an essential human resource for youth to draw upon in order to successfully navigate the different social fields in which they must participate.

Rowhea	If you don't have a squad, if you don't got boys to back you up, what would happen?
Shakeem	Dependin' on what type of person you is=
Rowhea	=What type of person you are.
Shakeem	Me. I'd be cool, because—hell yeah.
Rowhea	What about people like Randy?
Shakeem	He got mouth problems. It's like I got to let sh*t slide. I help people.

As pointed out by Shakeem, in the cases when an individual does not have a group to provide support, it is important to then find other ways to build solidarity and social capital, which may require knowing when to ignore a challenge to your status (disrespect) as well as actively seeking opportunities to help others.

Regardless of how it is accomplished, certainly, the building of social capital or the formation of tightly bonded relationships—for some, on the various levels of "associates, friends, and niggas"—can occupy much of the time of youth, extending into the classrooms of the comprehensive neighborhood school. Thus, considering the centrality of group membership to achieving and maintaining status and respect, and, sometimes, to warding off physical harm or ensuring an individual's safety, it is helpful to understand *playin* as the enactment of culture, belonging to another field yet appearing in the classroom, as a means for youth to build bonds of solidarity which will serve to garner and/or maintain social and symbolic capital in the classroom and out on the streets of the inner city.

Playin in the Neighborhood

As youth brought the camera into different spaces, to conduct self-ethnographic studies of their lives in fields other than the classroom, they often engaged in conversations with local youth, friends, and relatives. Of the many practices captured within the raw video footage, I identified a pattern of thin coherence, in which *playin* emerged as being unconsciously central to interactions in inner-city neighborhoods. Following the identification of various vignettes of *playin*, I began to engage in microanalysis to determine verbal and nonverbal cues that were indicative of whether emotional energy and solidarity emerged during the interaction.

Shakeem	About how long you stay in the mirror, man? Huh? How long you stay in the mirror?
Tom	I don't stay in the mirror. I ain't look in the mirror.
Shakeem	So how you know you're pretty?
Tom	I can just tell. Girls tell me.
Shakeem	Girls tell you!? Now I got to find these girls. Are they desperate?
Tom	Oh you're tryin to *play*? Ha, ha, ha! No, they ain't desperate.

Collins (1993) describes four main structural components of interaction rituals that contribute to the formation of positive feelings between individuals. Yet in the interaction above, the emotional energy was neither negative nor positive (i.e., flat) as Shakeem tried to *play* a boy, who had the reputation of being "pretty," by teasing him and ending with the sarcastic punch line, "Are they desperate?" Even though they were physically within the same place, bounded by some sense of boundaries, and mutually focused upon a discussion about Tom with the camera serving as a central artifact, Shakeem and Tom did not share a common mood nor did they have a mutual experience over which to emotionally connect. Rather, as Shakeem "tried" to *play*, there were breaks in the conversation flow with pauses longer than one second. There were no moments of overlapping speech where the end of one person's utterances was continued by the beginning of the other's statement. Throughout the interaction, Tom turned his head from side to side and even looked behind himself at times, although no one else was present. Moreover, even as Tom laughed in response to Shakeem's *playin*, there was a lack of synchrony with Shakeem, who did not join in laughter and seemed to struggle with a follow-up comment. As further explained by May, *playin* practices fail, as a method of building solidarity, when the person being *played* takes it the wrong way. Whereas Tom recognized Shakeem's attempt to *play* ("Oh, you're tryin to *play*?"), he provided a response that was emotionally flat, preventing Shakeem from being able to build any emotional momentum, thus the interaction faded into silence. Accordingly, *playin* practices are utilized most successfully (i.e., solidarity emerges) when interactions involve those individuals who embody shared understandings of *playin* and an unconscious "sense" or knowledge of the limits not to cross.

Both/And: *Playin* That Builds and Breaks Down Solidarity

Evident in the interaction between Shakeem and Tom which seemed devoid of emotional energy, is that p*layin* practices do not always result in the buildup of solidarity. In fact, although not the case in the above interchange, *playin* may also contribute to breaking down solidarity or, in some situations, these practices can accomplish both goals at once with the different individuals involved in the interaction. The following vignette provides an example of the dual nature of *playin* when Shakeem's disposition to *play* crossed the boundaries of the neighborhood and home fields and was enacted while employed as a student researcher in our project.

> On the morning of July 30, during one of the intensive summer sessions, we had planned to start off the day by continuing our discussions around social and cultural capital with the student researchers, but we were getting off to a slow start. Shakeem was sitting at one of our computer stations and holding the mouse. "What'll happen, right, if I took this Macintosh HD (hard drive icon) and drag it into the trash?" As he stated this question, Shakeem turned his head to the left to look first at one of the university researchers, Debbie, and then turned farther to look at me. I was sitting about two meters away at an angle behind Shakeem, and I leaned forward in my chair to look at his computer screen and focus on what he was asking. I had not heard him. "Shakeem, please

don't mess with it," Debbie requested. Shakeem's eye gaze shifted back to Debbie as she spoke; he smiled and turned back to the screen laughing. "Nah," he said." What is it?" I asked. Debbie was not amused. Speaking to Shakeem, she commented, "That's not even funny." Then in the same breath, she answered my question, "The hard drive." I responded in a questioning tone, "Oh, the hard drive?" At that instance, Shakeem turned again, this time it was not just his head but his entire upper body that turned toward me and he made eye contact. He started chuckling again, and, as I returned a smile, his chuckles broke into full laughter and drowned out Debbie's attempt ("ar-right!") to interrupt his *playin*. He turned a little bit further to glance at Randy who was sitting to my right. Randy's back was facing us since he was focused on the computer in front of him. Shakeem looked again at me, and made eye contact before turning back to his computer. He tried again.

Shakeem	That was so funny. I said, what'll happen if I took this (.) Macintosh [HD and dra:gged it to this -
Debbie	[I: don't even think it would let you. (.) But don't don't don't . .
Randy	(*He turned his head to his left in the direction of Shakeem who sat with his back to Randy.*) Well, try it and see!
Debbie	No!
Rowhea	He's just he's just [tryin' to get on your—(hh) he's not gonna do it. (*As I made this statement, I was smiling widely, understanding the playful nature of Shakeem's interaction. In response to the tone of Debbie's voice and my perception of her agitation, I raised my right hand twice making swift vertical movements gesturing for her to ignore Shakeem's playin.*)
Randy	[I'll try it.
Debbie	Randy, can you turn around? (.) Please?
Shakeem	It's right there! (.) (*As Shakeem paused, he looked over to Debbie with a serious expression and maintained his gaze for the next few seconds.*) All I got to do is let go! If I sneeze, it's over↑! You know . . . (*He turned his head back to the computer as he started to laugh.*) (hh) It's like, nah!

In this particular interaction, Shakeem made two attempts to *play* Debbie. Only recently introduced, she and Shakeem were in the beginning stages of establishing rapport and getting to know each other. In contrast, I had known Shakeem for several years and was well aware of his dispositions to *playin*. Yet, I did not become aware of his playful banter regarding the Mac hard drive until his eye contact, and bodily orientation toward me, engaged my attention. I then became attuned to the mutual focus of the discussion (i.e., Shakeem's computer). Even then, my reaction to Shakeem's *playin* was mild. My smile was small and I did not *play* back (mostly, because I did not know how); yet as Shakeem repeated himself for the second time, my smile widened and I laughed at various points, particularly as he made reference to sneezing. In fact, microanalysis reveals that my gestures to Debbie as well as my smiles and laughing indicate I have developed a "sense of the game" (Bourdieu, 1992) for Shakeem's *playin* practices and, hence, did not feel worried that he might follow through with his "threat" to delete the hard drive. Nonetheless, clearly, I had limited dispositions for responding to *playin* and for engaging in *playin* myself.

This vignette provides a means for studying more closely the ways in which *playin* can involve the use of language to give double entendre. In one sense, Shakeem's statement, "If I sneeze it's over!" suggests the literal meaning that the force of a sneeze could cause his hand to move, and he could accidentally drop the hard drive icon into the trash. In another sense, Shakeem uses the concept of a sneeze as a metaphor indicating to Debbie that no matter what she does, "it's all over." Just as a sneeze is difficult to control or stop, so too does Debbie have little control over what Shakeem will do—the hard drive will be deleted, no matter what. (Shakeem, personal communication, October 2003)

While some would find Shakeem's interactions as being resistant to participating in the research group's focus, here I argue that Shakeem was unconsciously enacting *playin* dispositions in an effort to build solidarity with Debbie, at least initially. His opening comments were directed to Debbie, yet as he was unsuccessful in engaging Debbie and experienced a lack of solidarity with her, Shakeem extended the conversation to include me, and then Randy. Interestingly, Shakeem even started over again to try again to build emotional energy around his comments. However, in this situation, as often is the case in classrooms, due to a dispositional disconnect, different ways of interpreting interactions and a lack of the "sense of the game," *playin* practices worked to eliminate Shakeem's opportunities to build a relationship with Debbie—although perhaps they simultaneously reinforced connections between him and me. As Debbie became audibly aggravated, Shakeem did not respond; instead, his focus shifted to trying to build solidarity with me and/or Randy.

> It's happened with me. It's like ar-right, you're *playin* too much. Calm down. Like, OK I understand we *playin*, but now it's time to stop . . . An' the line is so thin, you can never really tell when you crossed it unless you know who that person is—who you are hangin' out with. (Natalie, interview, 2002)

Knowing if a "line" has been crossed, or whether someone feels disrespected, is very important to prevent misunderstandings from arising through *playin* practices. During a summer session cogenerative dialogue, the student researchers spoke of the need to be attuned to verbal subtleties like tone of voice and oral cues during interactions. For example, the line has been crossed, they explained, "If you say 'chill' and they still doin' it." Similarly, when asked how an individual can unconsciously monitor interactions involving *playin*, Natalie also spoke of the importance of being attuned to verbal indicators, yet she additionally emphasized the role of nonverbal cues. Drawing upon a personal example involving one of her friends, Natalie explained, "She'll say something or she'll give a look." Thus, how far an individual actually carries on the *playin* is integral in whether or not the emotional energy built will be positive, negative, not there at all, or some gradient in between. While the interaction between Shakeem and Tom ended abruptly in the neighborhood, in the work-based vignette Shakeem drew out his *playin* longer than would be advisable (according to student researcher-defined limits) considering that Debbie was not *playin* back and had given him many verbal cues to stop. The comparison of these two vignettes suggests that Shakeem has a highly developed sense for knowing when to stop *playin* with peers but perhaps lacks knowledge of how to *play* with adults who do not share the same dispositions. Additionally, the computer vignette elucidates how easily *playin* practices could become misunderstood in a classroom setting and forefronts the importance of teachers developing a sense for when *playin* is occurring, not necessarily to *play* in return, but at least for the purpose of recognizing a student's gesture for building solidarity.

Playin in the Science Classroom

Playin practices structure and are structured by the field in which they are enacted. Moreover, they serve as resources for individuals as they exercise agency. Within a structure that does not support *playin* ideologically (i.e., work is a time to be serious), and which consists of human resources who do not have knowledge of the culture of others (i.e., you're being resistant), the agency of someone like Shakeem is rarely afforded by use of a disposition to *playin*. The possibility for structural resonance, or for aspects of structure to unconsciously elicit particular cultural enactment, is always present in any field due to the weak boundaries of culture (Sewell, 1999). However, when the structural conditions of a science classroom resonate such that *playin* practices (that are perceived as not "belonging") emerge, the repercussions include missed opportunities for building solidarity. In classrooms, the lack of such bonds between students and teacher and/or between students shapes the structure of the classroom field in ways that may detract from the development of a scientifically oriented community in which the doing of science becomes "fluent" (Tobin, chapter 2). In the following two sections, I provide two examples of *playin* practices within science learning contexts; the first is not associated with science activities and the second involves a science project.

Playin in the Science Classroom: Enacted Away from Science Activities

In Anita Abraham's chemistry classroom, her eleventh grade students shared many of the same dispositions, including those associated with *playin*. This became particularly evident on days when Anita provided them with a few minutes right at the end of class to unwind as she took care of tasks such as handing back papers.

> On March 20, 2002, the end of the period was fast approaching, and Anita stopped class a few minutes early before the students would head off to lunch to pass back graded homework assignments. As I had been a constant presence in this classroom for several months, the youth knew me well and I had come to have a clear sense of the tendency for the level of emotional energy to spike at times like this. On reviewing the end of this tape on this particular date, there was distinct evidence of high levels of positive emotional energy being generated between students. As a student spanned the room with the camera, it captured a large number of students out of their seats and walking around. While some of the students were handing back papers, others were up to socially interact. Regardless of whether the students were walking around for a "school related" purpose (i.e., handing out papers) or not, they engaged in similar practices, individually and as they interacted with each other. There was evidence of high levels of movement and verve (Boykin, 1986) as they moved in rhythmical ways, swaying while they walked and sometimes dancing alone or in synchrony with another. They sung and/or rapped to each other and to the camera, and there were many smiles and the constant sound of laughter as the youth teased one another and contested for the center stage. At certain points, the noise level would become loud such that Anita would "shhh" the class, yet even so, the emotional energy in the room remained charged, excitement was rampant, and students seemed genuinely happy as they had opportunities to connect with peers and join in practices which engaged dispositions to communalism, verve, movement, expressive individualism, and *playin*. In the midst of these interactions, one particular exchange captured my attention as it played through in real time. As Nathan came into the scope of the camera, he began to rap. He gestured for Marcy (who was handing back an assignment to a student sitting right behind him) to join him by putting his right arm around her shoulders to turn her toward the camera. As he rapped, Marcy's head began to vertically bob, in beat with his left extended arm which he moved up and down vertically. "Yo, my name is Nathan. I'm from the mountain. If you *playin* with me—I'm gonna slap you in your mouth." As Nathan let out the final line,

Marcy's head stopped moving as she loudly uttered a surprised "Ohhhhh!" with a wide smile upon her face. Within a second, Marcy replied to Nathan's rap by freestyling her own rap, "Yo. Yo. My name is Me. I don't like fleas so Nathan won't you get away from me?" As she began the rap, Nathan immediately began to sway horizontally from side to side. In resonance, Marcy also began to sway her body in rhythmic synchrony with Nathan. Remarkably, their interaction created a resonance within the classroom structure which afforded a third individual (approaching in the background on the left side of the screen) to join them, just as Marcy started to rap the sentence "Nathan won't you get away from me?" For three seconds, all three of the youth swayed together, moving from the right to the left in rhythmic harmony to the beat of Mary's rap. Yet, the interaction ended abruptly as Nathan realized that he had been *played* just as he had *played* Marcy with his own rap, seconds earlier. As Marcy gave the final line to her rap, she turned her face away from the camera toward Nathan and gestured to him with a strong, smug head nod. Nathan's face expressed real surprise; the third student smiled in the background.

What is significant in these sixteen seconds of interaction is the nature of what was being said and how, judged according to mainstream values, Nathan and Marcy could be viewed as taking shots at each other in rivalry. Rather, in contrast, they were building positive emotional energy around *playin* practices which were also combined with movement, verve, and expressive individualism. Such opportunities to build solidarity and community associated with being with peers within a science classroom should not be overlooked as menial in nature. There is much to learn from *playin* instances when students are not misunderstood or shut down, as was the case in this example from Anita's classroom. Certainly, even though not in association with science-related activities, these types of interactions are instrumental in helping to ascertain how solidarity develops between peers in science classrooms, what aspects of structure may contribute to the emergence of *playin* practices and how those practices then reshape the classroom structure. For example, recognizing the sheer enjoyment and synchronous nature of interactions where positive emotional energy is arising is key to understanding how such energy can be tapped so that the doing of science becomes a fluid process where "science knowledge, consisting of practices, facts, concepts, skills, interests, attitudes and values, is enacted without hesitancy, in timely and appropriate ways" (Tobin, chapter 2). I address these issues in greater detail in the final section of this chapter by focusing on the analysis of a vignette in which students are building positive emotional energy, solidarity, and fluency around doing science, as they engage *playin* practices in conjunction with a small science project during the first intense summer research session (2001). I identified this vignette, of less than two minutes duration, as salient for micro-analysis for three different reasons. First, it contained an example of *playin* between two student researchers, Kareem and Shakeem, and the graduate research assistant, Akram. Second, the interaction could easily have been interpreted as "off-task" behavior in a science classroom and, third, Shakeem's engagement with the science activity was not only uninterrupted but also enhanced, by the introduction of *playin*.

Studying Contradictions: Shakeem Focused on Doing Science

It was the end of July and I wanted to address concepts related to Newton's laws regarding forces, velocity, and acceleration. Drawing upon my elementary education background, I selected an activity I had done previously with students, the Puff Mobile (adapted from an AIMS curriculum unit). The activity required the youth to build their own cars and/or rockets, which were powered by balloons, from materials such as paper, straws, and Life-saver wheels. On this particular day, the camera was focused on Shakeem. In contrast to patterns which I had observed with Shakeem, both in his work with me as a student researcher as well as in his high school science classes, he was entrained upon building a frame for his vehicle. Shakeem sat at a round table with Randy (to his right), Kareem (to his left), and in front of him sat Akram, the research assistant. Rather than complaining about why he had to complete the activity, or putting his head down, as was the case on other days, Shakeem was attentive (for over an hour, he did not get up from his seat) and generally positive and upbeat in approaching the activity. The following transcript represents what occurred when Kareem began to narrate a story regarding how he had once stayed after school for detention and missed the detention bus.

Speaker	Time	Nonverbal Text	Oral Text
Kareem	00.00	*Shakeem's eyes are down and he is working.*	I had ta walk home one time, right? Cuz, I had detention and I missed the um the detention bus.
Akram	00.04.20		Ooh! °
Kareem	00.05.27		I had to walk all them miles, man.
Akram	00.07.29		*Laughs hissingly through his teeth.*
Randy	00.08.12		What [did your mom say?
Akram	00.08.15	*Shakeem's shoulders*	[All them miles!

Shakeem	00.08.15	*and face shift slightly to the right, toward Kareem. His eyes are not visible. He is smiling.*	[All them miles!
Shakeem	00.09.25	*Shakeem's face turns back to the front. His eyes are down on the design in front of him.*	
Akram	00.10.01	*Shakeem's eyes remain down and he is working.*	With one shoe!
Akram	00.11.08	*Shakeem breaks out into a wide smile with his teeth showing. He leans slightly forward as he laughs, and his shoulders bob up and down. His hands shake from laughing as he cuts a piece of tape.*	*Laughs sharply and loudly.*

In this interchange, the talk was not scientific (i.e., the discussion did not address either aspects of the student researchers' car designs or any concepts related to the motion of the vehicles). However, in viewing the video segment without the accompanying audio, what is visible is a student thoroughly engaged in the task before him, working meticulously to tape together four plastic straws to serve as a framework and as axles to hold lifesaver wheels. Shakeem held a smile on his face as he worked, which really widened as he gave his *playin* response to Kareem's narrative. However, even as he *played* Kareem, his focus remained on his car design with a shift in attention for only about one second in duration. Also salient is Shakeem's synchrony with Akram as they overlapped in exact harmony ("All them miles!"). Although Randy also overlapped in almost concurrent timing, he asked a question rather than engaging in the *playin*, suggesting that, in this situation, Shakeem and Akram were in sync with each other but not necessarily with Randy. Without a break in rhythm, Akram continued to *play* Kareem by adding on the comment, "With one shoe!" Shakeem's nonverbal response occurred in synchrony with Akram's laugh, although this time, his eye gaze and facial orientation did not shift at all. Instead, Shakeem picked up the scissors and cut a piece of tape for his design, even in the midst of his laughter. Here microanalysis suggests that the synchronous nature of Shakeem and Akram's verbal utterances and nonverbal gestures, as they engaged *playin* practices, contributed to the generation of positive emotional energy and feelings of solidarity, particularly for Shakeem, which continued to build as the conversation unfolds.

This ten-second interval provides evidence of Shakeem's simultaneous entrainment with both the conversation with Kareem and Akram, as well as with his task of building an air-powered car. Although his eyes were focused upon his design, Shakeem was able to complete Akram's sentence ("Yeah, but if you walked home=" "=You didn't get home 'til six, nigger.") with only a 00.00.06 interval between their individual utterances. Interestingly, when Shakeem did glance up (beginning at 00.27.17), it was to Akram, rather than Kareem although Shakeem had just made a statement directed to him ("You didn't get home 'til six, nigger"). Then, perhaps in response to Shakeem's glance upwards, Akram continued the conversation by restating (at 00.27.22) their mutually expressed *playin* sarcasm ("All them miles"). Akram's statement and light laugh resonated with Shakeem, who then smiled.

While the science activity I initiated was structured such that each student was individually focused upon his own design thereby encouraging competition versus collaboration, the social talk emerging, and particularly Kareem's narrative, provided an opportunity for feelings of solidarity to develop among the

Speaker	Time	Nonverbal Text	Oral Text
Akram	00.17.07	*Shakeem's eyes are down and he is*	You must have gotten home pretty late, huh?
Kareem	00.19.06	*working.*	Nah°,we got out of school 2 - 3:30 and detention was over at 4:30.
Shakeem	00.22.22 - 00.23.11	*Shakeem's face shifts slightly toward Kareem and then back down to his project.*	
Akram	00.25.05 - 00.26.05	*Shakeem's eyes are down and he is*	Yeah, but if you walked home=
Shakeem	00.26.11- 00.27.12	*working.*	=You didn't get home 'til six, nigger.
Shakeem	00.27.17 - 00.28.01	*Shakeem glances up to look at Akram and then back down to his work.*	
Akram	00.27.22	*(Shakeem's eyes are down and he smiles as he works.*	=All them, all them miles (hh).

group members. First, casually sitting together at one small round table and working on the same science-related goals, the physical structure of the learning environment invited Shakeem, Akram, Kareem, and Randy to engage in social conversations in addition to science talk. Second, ideologically, since the youth were employed as student researchers, this learning context was structured by a rule system in which student talk was not being monitored, as is often the case in schools. Thus, arguably, the schematic and material aspects of this field's structure supported the emergence of Kareem's narrative as a key resource for affording Akram's and Shakeem's unconscious enactment of *playin* practices, which in turn served to restructure the learning environment to foster solidarity rather than competition. Around the mutual focus of the narrative, positive emotional energy emerged, evident in Shakeem's smiles and laughter, as he and Akram shared an experience (i.e., overlapping synchronous thoughts of how to *play* Kareem) that contributed to Shakeem remaining deeply entrained in the project. When he did become "distracted" from his individual design, it was to follow a science-oriented discussion about Kareem's design.

Speaker	Time	Nonverbal Text	Oral Text
Akram	00.49.21		Kareem, you ain't got an eighteen wheeler!
Shakeem	00.50.13-00.56.03	*Shakeem's eyes come off his car and he looks at Kareem's design.*	
Kareem	00.50.21		Nah, but see, if I don't have a lot of these, it will go up because of the weight from the balloon.

For the first time in the fifty seconds that elapsed since the conversation with Kareem began, the talk turned to science and, only then, did Shakeem's focus move from his own design. Actually, Shakeem's eye gaze remained on Kareem's car for six seconds (00.50.13-00.56.03), while Kareem justified his design decision to Akram. Outstanding in this portion of the clip is the distinct difference in attention (six seconds versus one second) that Shakeem gave to Kareem's science-related comment, in comparison to the prior socially relevant dialogue. This signifies that social conversations, interlaced with *playin*, do not necessarily detract from students' attention on the science activity and, may instead, contribute in unassuming ways to the generation of positive emotional energy and solidarity around the experience of science.

Akram's role within this science activity is also very significant to consider. His embodiment of dispositions to *playin* resulted in a very different type of interaction than what occurred in the vignette with Shakeem and Debbie. The latter vignette was markedly fragmented and laden with misunderstandings and discomfort amongst those involved. In fact, *playin* was viewed as a practice that needed to come to a halt in order for the planned theoretical discussion to begin. In contrast, the interaction between Shakeem, Akram, and Kareem was distinctly fluid, in that utterances, including those that involved *playin*, flowed seamlessly, as a conversation evolved around the doing of science. Akram's presence in the group and his engagement in *playin* with Shakeem and Kareem further afforded Shakeem's agency as a science learner, and the making and (re)making (Roth, Tobin, Elmesky, Carambo, McKnight, and Beers, 2004) of his identity in science, in terms of being able to problem solve and understand concepts related to forces, air flow, and acceleration. For example, later in the morning when Shakeem ran into challenges in his design, rather than giving up or expressing negative emotions in relation to science (as he often did), Shakeem spoke positively about himself and engaged in discourse to decide on which side of the car he should place the balloons to propel it forward.

> If I put the balloon here, where the hell is the air going to come from? The pen's stuck in the front. Bad idea. I can eat the candy now. Start over. (*Shakeem pretends he's going to eat the lifesaver wheels.*) It's like "No!" How can I work around this? Cuz I like the pens. (*Chuckles*) I love myself. Damn, I'm smart, man! (*Shakeem takes his frame and cuts off the parts of the straws that are not part of the car structure. He takes two new straws and positions them tentatively in new positions on his car frame.*) Wait a minute. Wait a minute. Air's gonna be comin' this—damn it. How did that work out, man? Yo, A [Akram]. Air is—if air is goin' in this way, it's gonna push the car (0.2) if the air is goin' out this way, it's gonna push the car this way?

Moreover by studying, on a micro level, Shakeem's practices to enact *playin* dispositions during this science activity, we have gained insight into how he builds solidarity around being involved in science and how that may impact his development of a scientific identity. Clearly, having the opportunity to participate in a science community, without having to constantly focus on monitoring unconscious *playin* dispositions, can be extremely empowering to students like Shakeem who are often squelched within traditional science classroom structures.

Coda

Rowhea	Do you think having the teachers understand this way of -
Natalie	=that it's all out of *play*?
Rowhea	And *playing* back a little, does it=
Natalie	=It does make the class more enjoyable. I knew it was coming. It helps move the class along. It makes the class more bearable.

For many students, and many teachers, inner-city classroom experiences are hardly "bearable." Teachers have lost hope in being able to reach youth who they consider far removed from society's academic and social standards, and students have lost hope in education as a means for transforming their lives. When asked about their aspirations and dreams, the students speak of great achievements—among them, family, riches, and fame. Yet, these dreams are, more often then not, tied to excelling in sports, acting, or the music industry, and rarely to science. As shown in this chapter, resolving some of this alienation from school science requires us to forefront the significant ways that dispositional disconnects and varying sociocultural histories shape what happens in the classroom.

The majority of youth attending comprehensive neighborhood schools like City High are being taught by teachers with whom they cannot build working relationships, much less any sense of real solidarity. While recruiting new teachers like Akram, who understand the many different practices that inner-city youth may bring with them to the classroom, seems logical, it is far from a solution to the challenges of urban science education. Rather, it is what we can learn through studying teachers like Akram and their interactions with youth like Shakeem that we can come to better understand how to improve the teaching and learning of science in urban schools. In actuality, what we need are more teachers who are able to develop an innate "sense of the game" for student practices like *playin* and who understand these practices as being important resources for building social and symbolic capital and solidarity for many inner-city youth. Moreover, this research also suggests that teachers develop a deep conceptual grasp of the both/and nature of such practices so that they can come to view *playin* not as destructive versus constructive but as being both/and simultaneously for different people involved in the interaction. It is also important for teachers to consider the ways in which *playin* practices are self-regulated by youth who are very aware of subtle nonverbal and verbal cues (mostly unconscious and rarely articulated) that indicate when a line has been crossed. This knowledge can help teachers feel less anxious about the need to control the unfolding *playin* practices in their classrooms. Incorporating student voice in making sense of these variations of what is happening in science classrooms is invaluably accomplished through cogenerative dialogues, and this is demonstrated in powerful ways in chapter 8.

This research reveals that dispositions to *playin* can shape the structure of a classroom, since as they are introduced, they become resources around which mutual focus can emerge, emotional energy can be built, and the development of solidarity can occur. Thus *playin* practices have the potential for increasing student participation, generating excitement around doing science, and contributing to the creation of science communities. Moreover, as *playin* contributes to the development of social networks, it becomes an important tool to aid students in accessing and appropriating other resources in the science class. Through such understandings, I urge us to consider a more flexible definition of science community, as well as what we consider to be science discourse, to include the seamless enactment of verbal *playin* practices while "doing" science.

Chapter 6

All My Life I Been Po': Oral Fluency as a Resource for Science Teaching and Learning

Gale Seiler

All my life I been po'
But it really don't matter no mo'
And they wonder why we act this way
Nappy boys gon be OK

Although written and performed by Nappy Roots, a hip-hop recording group from Kentucky, the above lyrics also describe life for many African American young people in inner-city schools. They are members of an underclass who live in segregated, economically disadvantaged neighborhoods and for whom there is little chance of social and economic mobility (Wilson, 1996). One might wonder what the lyricist means when he says, "Nappy boys gon be OK." For African American students who live in poverty in American cities such as Philadelphia, it is unlikely that these words mean that their education will enable them to change their position in social space and move out of the underclass. Statistically they are more likely to go to prison or drop out of school than they are to go to college. Overall schools have been ineffective in changing the life courses of such disadvantaged students. I am convinced that one of the reasons for the inability of education to bring about such change is that, as teachers of inner-city students, most of us lack something that is needed. Murrell describes what is missing as a "deep-seated understanding of African American experience, culture, and heritage and the ways that this understanding informs successful teaching of African American children" (2001, xxiii). Thus we continue to wonder why they act "*that*" way and prefer that they act in other ways. If instead, the students' ways were an accepted and valued part of classroom practices, related structure could be changed and the agency of students and teachers enhanced.

I, and others, have written about the need for education, particularly science education, that is transformative for African American students (Barton and Tobin, 2001; Seiler, 2002). But what do we mean by transformative? On a macro scale, we want marginalized students to have opportunities to exercise choice in their life trajectories and social positions. In terms of school-related agency, we would like them to acquire knowledge, skills, and dispositions that can help them meet their own goals and be able to participate in professional discourse and science discourse if they choose. In that sense, the liberation rests with the student (individual or micro level) as well as in his or her interactions over time with other students and teachers (meso level). However I also believe the transformation can move outward to the macro level. This involves changing the structure of academic discourses, as well as scientific ways of thinking and being, that work counter to the agency of students who are already marginalized in society. The three levels (macro, meso, and micro) represent three foci of inquiry that can direct our research and understanding of this issue of social transformation, and they can also guide us in attempting to bring about change in one level via transformation at other levels, to the extent that the levels are transactionally linked and responsive to each other. Changes in the practices of teachers and students in the classroom can be initiated and studied across both micro and meso levels. These changes can then effect change in structure and agency on macro levels so that cycles of reproduction can be broken. As conceptualized in the introductory chapters in this book, the dialectical relationship between structure and agency connotes that changing aspects of either, has the potential for changing the other, in a recursive fashion. Accordingly, the agency of participants depends upon the existing structure, but the practices and schema associated with agency can transform structure and change the potential for action of all participants.

This chapter is intentionally focused on the positive practices and dispositions of the African American students at City High. This focus is not intended to paint an overly rosy picture of the students' behavior or their prospects in school or life. Rather it is an attempt to counter the negative stereotypes commonly associated with inner city African American teens and to encourage people from other lifeworlds to appreciate the rich cultural resources that these students offer.

In addition, I focus on goals that schools might help students meet. However I do not minimize the presence of other significant obstacles faced by African American students in the unjust and racist society in which we live.

Connecting the Curriculum with Student Interests: "Hello Mr. Sheepie-poo"

An academic achievement gap has been described between black and white students, regardless of socioeconomic status and other considerations, but the gap is most profound when considering inner-city African American students. Many ideas have been put forth to attempt to understand, explain, and in some cases rectify the factors involved in this continued achievement gap. A frequently heard explanation regarding the gap is related to the lack of connection between schools and the cultural and linguistic experiences of African American learners. One approach to this lack of connection is to create greater student engagement by connecting the curriculum to the lifeworlds of the students.

Recently Delpit (2002) described connecting middle school curricula with the interests of African American students through the study of hair and hair products. We have used similar approaches for several years and have employed and involved high school students as curriculum designers and teacher educators to help us understand how curricula framed around student interests might look. In a science lunch group (Seiler, 2001) and as summer employees, high school students have pursued ways to help us connect science with interests of theirs such as drumming and rapping (physics of sound, anatomy of the ear and vocal chords), and the World Wrestling Federation (physics of motion and force, and anatomy of human joints). In an effort to give students a voice in the science curriculum, a biology elective course was taught at City High in which two coteachers kept an eye on science standards while responding to student suggestions for specific topics and learning approaches in what we called a student-emergent curriculum (Seiler, 2002).

The science lunch group and the biology elective both provided evidence of the power of foregrounding student interest and putting curricular choice in the hands of students, two approaches not traditionally found in schools. In both of these situations, structural modifications afforded student agency and precipitated other structural transformations. During the biology elective, a compelling student interest emerged in dissections; the students repeatedly asked to do dissections and suggested organs and organisms that they wished to dissect. The extant structure, composed of material resources (e.g., scalpels) and classroom practices, rules, and schema provided constraints that initially made the coteachers reluctant to do dissections. No science class at City High had done dissections in a number of years. However, the coteachers eventually obtained specimens to be dissected and audio and videotapes from the class periods when dissections were carried out show that these sessions were unique in several ways. These differences provide evidence of the transformative potential of changes in structure and practices.

A cultural toolkit consists of a set of symbols, meanings, and social skills that serve as cultural tools from which actors construct actions and practices (Swidler, 1986). These are derived from the cultures and subcultures in which a person participates. The resources can be used creatively, combined in new ways, and used in new settings but ultimately our lines of action are constrained by what is in our toolkit. When listening to and watching the tapes of sessions involving dissections, there were marked differences in the level and type of student engagement and the tools the students used compared with other days in the class. The students were more focused and attentive and used their cultural toolkits as adeptly as they did their dissection toolkits. In the following excerpt Cedric and Shakeem, two students in the biology elective, are dissecting a sheep skull.

Cedric	There go his tongue right there. (*Pointing to part of the sheep head with a probe.*)
Shakeem	Mr. McGow, check this one out. Look, John. Here go his upper jaw and his lower jaw.
McG	Uh huh.
Shakeem	His tongue. His nasal cavity right there.
Cedric	Wait, wait. Where the nasal cavity at? This part right there?
Shakeem	Yeah. And here go his eye sockets.
Cedric	Here the ear holes right here.
Shakeem	Oh snap, man.
Cedric	Yo, where the spinal cord at? Back here? Where the orbits at?
Shakeem	They the eye sockets.
Cedric	So everything alright on this one? (*Pointing to the drawing he had just labeled on the handout.*) Where the adipose tissue?
Shakeem	Hello Mr. Sheepie-Poo. You know there's a sheep Pokemon?
Cedric	It say describe the adipose tissue. (*Referring to the handout.*)
Shakeem	Where that?

Students at City High commonly engage in a variety of behaviors that seem to postpone or avoid doing work in class. These actions are so prevalent in this setting that we have come to call them the urban shuffle. These patterns seem to be due to the enactment of schema, resources, and practices from the street (Anderson, 1999) in the classroom and may be connected with a disposition toward a social time perspective (Boykin, 1986) in which time is treated as a social space rather than a material one. On most days in this class, the urban shuffle was prevalent. Off-topic comments and behavior were more common than science-related comments and actions and they often continued for lengthy stretches. However the above transcript illustrates and video analysis supports the relative absence of off-topic talk and actions compared with most class periods. Shakeem's off-topic utterance about Mr. Sheepie-poo and Pokemon shows his use of humor and association with popular culture but his comment was not pursued and the pair quickly returned to their task. Shakeem and Cedric exhibited great pride in what they were doing when they called Mr. McGow over and identified the parts of the skull for him. The pair worked in tandem, going back and forth, to identify the parts for themselves and for Mr. McGow. They used their own way of talking, which included slang and black English and was not very science-like in a standard sense, yet they worked together to identify the parts and label the diagram appropriating new vocabulary as they proceeded. What is not discernable from the transcript is the sound of their discourse while carrying out the dissection. The cadence, rhythm, and feel of their talk was that of inner-city black male teenagers and if overheard, might be taken to be off-task or inappropriate for science class. These individual speech patterns in addition to the meso level pattern of exchanges between the students can easily get in the teacher's way and lead to macro patterns and practices in which schools disadvantage underclass students. The involvement of students in practices using their own ways of talking and being often results in shut downs by the teacher (Tobin, chapter 2) rather than their use as capital through which to advance learning.

When listening to the tapes it was startling to hear the extent of the orality of the students while carrying out the dissection. They read the directions and questions aloud, spoke their answers as they wrote them, identified anatomical parts aloud, and sang to the sheep. Taking a line from a popular R and B song by Musiq Soulchild, Shakeem improvised changing "I just wanna know your name and maybe sometime we can hook up, hang out, just chill" to "I just wanna know your name and maybe sometime we can cut up, this brain, with scalpels." This disposition for orality and joint activity highlights the importance of the creation of time, space, and opportunity for the social construction of knowledge in classrooms with all students, but particularly with African American students. These dispositions represent cultural capital that students offer for use in science teaching and learning.

As described in the introductory chapters, structure and agency exist in a dialectical relationship and this offers the possibility of reproduction of existing structure as well as production of new structure and culture. Therein lies the potential for change, as in this example where teacher and student goals were both met. The coteachers were cautioned against allowing these particular students access to scalpels, or investing financial resources in the purchase of specimens to dissect. The expectation was that the students would not approach it seriously and would learn little. It was expected that at the least, they would "just screw around" and at the worst, they would get into trouble with the scalpels. In choosing to do dissections, the coteachers altered classroom structure by loosening certain constrictions on behavior, countering the expectation of misbehavior, and creating new structures that placed the students in positions of responsibility. What we do in schools, how teachers and administrators from mainstream culture anticipate and respond to students of color, emanates from and represents how the larger society reads these students. By responding negatively to certain aspects of the students' culture, the oppressive structure is reproduced in the school. The recognition that structure is changeable and continually re-formed creates possibility and gives all participants collective responsibility for its formation and transformation.

The students exhibited great care in their work during the dissections and the urban shuffle was much less evident. They followed the directions on their handout fastidiously and carried out precise dissections of the brain. Their eagerness counters our stereotypes of inner-city African American high school students who are commonly seen as disinterested in and disengaged from science. At times the pairs of students sounded like surgical teams as they worked in unison to carefully remove the fragile brain from its cranial case.

Cedric	Lemme lift it up. Then you can do it like that. No. Across there cut it. Yeah.
Jen	I would take the forceps and
Cedric	Hey why don't you cut that off?
Shakeem	I swear to God I seen some juicy stuff.
Cedric	Put it under there like that. Shakeem we gotta get this part.
Shakeem	Hold that down there.
Cedric	You should cut that.
Jen	Flip it over.
Cedric	Alright, but hold it. Let us get a little more of this out.

Here Cedric and Shakeem acknowledged the teacher's (Jen) suggestion, but worked autonomously to carry out the dissection of the brain. They exhibited great cooperation and coordination of their movements as they offered and accepted suggestions, exemplifying the communalism that Boykin (1986) talks about. They did not abandon their cultural toolkits while doing dissections, rather they used them in creative ways to gain respect from their peers and adults and to complete their work. Something about doing dissections triggered the students to draw more selectively from the actions and practices they had available in their cultural toolkits. Months later Keisha, a female student in the class, said she hated nearly everything except the dissections: "Well when we was doing that dissection stuff. I liked that. That's the only thing I liked." Was it that they were finally getting the chance to do something they had requested? Was it that access to dissecting equipment positioned them in roles of trust and responsibility? Was it that they gained respect and social capital from Mr. McGow and visitors from other classrooms? The dissections were a wild success; something different happened in class on the days when dissections were done.

Like Delpit, I have seen the power of linking curricular topics to student interests and providing opportunities for student voice in curricular decisions. There are hundreds of topics with which this can be done, and an endless number of connections that can be made to even a single topic or activity, such as playing basketball, baseball, video games, and drums, or cutting hair, recording music, and the science seen in movies (Seiler, 2001). Moreover, this can be done in ways that address science content and process standards. Attention to students' topics or questions can contribute to student agency and lead to the emergence of new patterns of practice to some extent, but changing the focus of science content is not enough. We must find ways to recognize capital in less obvious aspects of students' culture so that agency might afford learning in more significant ways.

Culture of Power

Although educators have moved away from a deficit view and have become more willing to consider a variety of factors in understanding the poor school achievement of African American students, school personnel still hold both spoken and unspoken expectations of white, mainstream ways of being and monochromatic norms of student participation. If we adhere to the current structure in the form of rules and schema, we insure that schools will continue to play gatekeeping roles and that society at large will continue to discriminate against African Americans mired in an underclass. Many in education recognize the importance of opening the gateways that have hindered blacks and other minorities in school. One way that has been suggested to do this is to explicitly teach students the norms and conventions of Standard English while at the same time to honor the language of the students (Black English). Delpit notes that society unfairly holds people accountable for a set of rules about which they have never been directly informed (1988). In her writings, Delpit refers primarily to the acquisition of Standard English and calls for teachers to make explicit its stated and unstated rules. She provides examples such as teaching the conventions of print and oral communication, the technical rules of writing, and Standard English grammar. She also stresses the importance of enabling students to appreciate the value of their personal code and the dynamics of power in this country.

Delpit refers to the disciplines of reading and writing and focuses largely, though not exclusively, on primary grades and adult learners. She remarks that her suggestions refer best to those who share a "core black culture" and not to those who are participants in the marginalized underclass. I have read Delpit's writings many times and considered their implications for underclass teenagers in inner-city high school science classes, that is, for the students I have come to know in Philadelphia. I suspect that, for these particular students, their transformation and potential success in school and in life, is tied to something more than language, more than teaching them the rules and conventions of written and spoken Standard English. While I am not downplaying the need to learn to read and write in ways that afford success, I recognize that many of the students at City High are already quite adept at code-switching from and to Standard English. Thus it seems that we need to think beyond language to a holistic view that includes cultural styles and attributes that are embodied in the students' actions and ways of being. These styles and attributes are part of the schema and practices that interact dialectically to shape structure and agency in school and in life.

Boykin (1986) suggests that black culture is in opposition to the culture of mainstream America. He writes, "To characterize Afro-Americans as culturally different from Euro-Americans is not graphic enough. To the extent that the black experience reflects a traditional West African cultural ethos, the two frames of reference are noncommensurable. There are fundamental incompatibilities between them" (63). Boykin enumerates nine dimensions of African American culture emerging from its roots in West African culture, and these attributes are frequently expressed by students in school, but are rarely seen as resources through which agency might be created and structure transformed. In the previous examples during dissections Shakeem and Cedric used their sense of communalism, orality, and rhythm, and there are other dispositions as well that can be used in ways that are agentic and can transform structure, if agency is not truncated.

Enacting Culture from Their Toolkit

The African American students at City High display amazing verbal abilities. Many are wonderful mimics, comically imitating Ken Tobin's Australian accent, the speech of Korean shop owners in the neighborhoods of West Philadelphia, or the speech of white suburban female teens. They have the ability not only to "pick up" new ways of speaking, but they also possess an awareness of when and how to use them, a metalinguistic ability that is quite remarkable. They "talk proper" when they use hypercorrect English and exaggerate the acquisition of new science vocabulary, as when they began to ask to use the bathroom in a number of new ways, after learning about the excretory system. Taking on an exaggerated pronunciation and playing with words to make a phrase such as "a longevity of urine" illustrate how the students often combine their verbal dexterity with humor, sarcasm, and scientific terms. Their ability to argue and use evidence to substantiate their claims can be turned from debating the best player in the NBA to the reason a cockroach can survive a fall from eight feet or more. But these verbal exchanges often contain elements of signifying, a genre of talk involving indirection and double entendre, and requiring participants to look below surface meanings and make inferences. "Playin the dozens" is one particular form of signifying. Signifying often involves ritual insult and satire or "woofin," a verbal bluff or threat and thus can approach proportions that seem like verbal combat (Smith, 2002). The inferential thinking involved in these types of speech enriches both the everyday and classroom speech of African American students with metaphors and analogies. However in school, these abilities that are embedded in black culture often lead to discomfort and shut downs on the part of teachers unfamiliar with them.

Thus far few educators have begun to see these cultural dispositions as resources for learning and building understanding in specific disciplines. An exception is Lee (1992) who uses the students' figurative speech abilities related to rap music to enable them to interpret and construct understandings of mainstream literary pieces. I have found that many of the examples, comments, and questions that students offer in science class represent valid efforts to connect science to their lives and can be useful in advancing individual learning and, when offered publicly, advancing the learning of the group. For this to happen, space and opportunity must be created by changes in the classroom structure, schema, and practices of the science classroom.

The following examples illustrate the eagerness with which African American students at City High dip into their cultural toolkits and use their experiences and cultural styles to talk and think about science. In the excerpt below, one of the coteachers (Ryan) was leading the class in a discussion of heart rate and factors that affect it.

Ghetto Analogies

Nicole	When I walking home and get scared.
Shakeem	If somebody behind you, you sweat.
Ryan	Did you ever hear of an adrenalin rush? You know what adrenalin is?
Shakeem	Yeah, it's like a woody all over your body.
Keisha	Eminem said that.
Shakeem	Like Popeye on spinach.

Shakeem coined the phrase "ghetto analogies" to denote the many figures of speech that African American students create and use in and out of school. His use of the referent "ghetto" denotes that this figurative language draws from the students' lifeworlds that are different from, and likely not understood by, their teachers. In science class, their ghetto analogies are evidence of underlying reasoning and sense-making that is worthy of attention. As shown in this segment, students frequently offer connections between science and their lived experiences, in this case a racing heart and sweat in response to particular occurrences on the street. Shakeem is a master of analogies in his speech. He vividly compares an adrenaline rush to an erection "all over your body" and to Popeye's response to spinach. These statements represent two different aspects of his culture—a sixteen-year-old's tendency to see sex in everything and a cartoon character. Shakeem's flair in using analogies illustrates his historical connection with "*playin* the dozens" or signifying. Whether rapping or talking, Shakeem's use of figurative speech and other oral abilities are rapid, precise, and clever. However on this day his analogies were, as often is the case, lost amidst the other sounds and activities in the class or perhaps they were ignored by the teacher as inappropriate for classroom talk. The creativity and possibility of such analogies were only recognized later in the video and audiotapes; they were lost in the unfolding classroom events at the time or perhaps unacknowledged by teachers with different sensibilities related to cultural and age differences.

Student-generated analogies represent attempts to connect new information with prior experiences both in and out of school and to connect the unfamiliar with the familiar in their lives (Wong, 1993). The nontraditional comparisons demonstrated in student analogies can serve multiple purposes and can be powerful tools in teaching and learning. Analogies can make observations and attributes accessible to others. What features of a "woody" or Popeye or spinach led the student to associate them with an adrenalin rush? Are those associations reasonable? Analogies can also lead to the analysis of possible functional relationships. What is the relationship of spinach to Popeye and what plays the role of spinach in an adrenalin rush? Expressing an analogy orally and putting it into the public sphere in the classroom can serve as a basis for communication about the object or concept. The construction of such a figure of speech provides a link between the students' lifeworlds and the science classroom, and a link between the micro/individual level and the meso level where learning occurs. Student thinking as glimpsed through their analogies emerges from their experiences and cultural histories and can engage the students and teachers across time and space in creating new classroom practices. This creates the possibility for macro extensions and the evolution of new norms and practices.

The propensity for bringing their own experiences and examples into the classroom from fields outside represents agency on the part of the students. In doing so, they alter the structure of participation and engagement in the science class. The occurrence of student to student cross talk in many excerpts from the science lunch group and the biology elective class (where student knowledge was centered) counters the more common but oppressive teacher-directed patterns of discourse in science classrooms (Lemke, 1990). Agentic behavior on the part of the students altered some aspects of the classroom structure and culture, but these are eventful classrooms with beginning teachers. The possibility of building on this to further advance learning was fleeting. Without immediate recognition (either conscious or unconscious) and action by the coteachers, these moments usually passed with their potential unrealized. A crucial question emerges—how can we push student agency into a central position so that it can afford learning and become a resource for teaching?

By placing more emphasis on what the students already know (e.g., experiences with physical manifestations of fear) and can do (e.g., the use of figurative speech), we may be able to connect the culture of power with the students' culture so that it supports learning. Sewell's definition of culture (1999) tells us that the boundaries between cultures are porous and weak. We see this whenever students use their schema and practices in science class, even though other ways of being are usually privileged. By allowing students' cultural attributes to be foundational to the development of new practices and norms of participation in the classroom, the porosity of cultural borders would become an asset for the students and a resource for teaching and learning, instead of a source of friction between students and teachers and failure for the students.

Nonverbal Expressiveness

The text-based nature of this book precludes the inclusion of many instances in which students employed their sense of verve, rhythm, and communalism in nonverbal ways while participating in science. Through the following excerpt I attempt to show their expressiveness beyond the use of words. In this excerpt the students and the teacher, Jen, are talking about the control of unconscious processes such as breathing by the nervous system.

Jen	How do you keep breathing while you're sleeping?
Shakeem	Yeah, that's a good question, yo. 'Cause your brain be knocked out. You be like *zzzzzz*. (*Making a snoring sound.*)
Pedro	Your brain's still at work.
Shakeem	Your brain on layaway.
Jen	Like you're not awake so how does that all work?
Shakeem	'Cause your brain keep one eye open. Your brain be like. (*Imitating it with one eye open and one closed.*) Everything else be chillin and you be breathin. (*Making a panting sound.*)
Nicole	That boy is stupid.
Shakeem	See cause ya'll don't think about stuff like that. Getchur minds right, all a y'all.

Here Shakeem used sound effects by imitating the sound of a person snoring and panting. With his facial expressions he offered an impression of the brain with "one eye open." Shakeem's analogy shows his ideas about how the brain controls breathing by drawing on a common cultural phenomenon, putting a purchase on layaway. But how do these figures of speech, the idea of a brain on layaway or with one eye open, relate to the control of breathing by the autonomic nervous system? What attributes and relationships were used by Shakeem in his analogical reasoning? Perhaps related to the hint of an altercation between Nicole and Shakeem, these analogies remained unexplored in the class that day, though the teacher was sensitive to and struggled to respond to student input. Again student agency re-structured the classroom discourse patterns to be student-centered, yet this transformation of structure passed and did not become a resource for learning or for further transformation within the class.

Social and Cultural Histories

A visitor to an eventful classroom at City High School or other inner-city high school, if asked to identify what the "problem" was, would probably refer to the "uncooperative, disinterested behavior" of the students. Such students and their behaviors are often labeled as resistant. I no longer use that term since it locates the problem in the actions of the learners and does little to enable us to affect change (Seiler, Tobin, and Sokolic, 2003). Cultural Historical Activity Theory (CHAT) (Engeström, 1999) helps us to understand that the system of schooling practices fosters behavior seen as resistance, disengagement, or lack of motivation. The system of practices fails to appropriately address the social, cultural, and historical context of the schooling experience of African American students and does not recognize the need for classrooms and schools to be designed in ways that multiple goals of students and teachers can be met. As shown in the excerpts included here, a classroom system is not controlled by the teacher but rather is generated in the interaction between teacher practice (both planned and in-the-moment), the activity of the students, the various tools and artifacts used in class, and the participants' goals. Outcomes emerge from these components; yet, unfortunately, for most inner-city science classes, significant learning is not a frequent result. The dissection activities represent a rare alignment of goals in the biology elective class in which the student and teacher actions provided opportunities for the teachers' goals (student learning and a less chaotic classroom) and the students' goals (gaining respect and studying a topic of interest) to both be met. Recognizing the interplay of all these components takes us beyond student behavior problems and culturally relevant curricula, to a perspective of learning that is situated and emerging *in* this time and place yet emerging *from* the social and cultural histories of the participants, both students and teachers.

On a daily basis, students make connections between science topics and their own lives and experiences. Many times they use these experiences to pose interesting questions related to science. In the biology elective class Derek asked, "How come somebody I knew got shot in the leg eleven times and he didn't know it 'til he sat down and saw the blood? They were shootin at him and he was still out on the street and he still runnin. He said he didn't feel nothin touch him 'til he got to his girl house and he sat down on the steps and he saw all this blood leakin' down." The content and style of such talk often leads teachers to form assumptions about students' seriousness and commitment to learning. As other students contributed additional stories of being stabbed, and raised more questions about the ability of a person to block out pain, the teachers' discomfort level rose. It is unfortunate but not surprising that the coteachers were unable to see the possibility for new practices in the students' questions and stories, since teachers' sensibilities and expectations emerge from life experiences that are quite different from those of the students.

In the following interaction, Shakeem refers to the food provided in the cafeteria under the federal free lunch program as the "freebie." He considers it in light of an issue that had been in the news that year—the presence of high amounts of lead in the water of many Philadelphia schools including City High.

Shakeem	I got a question, yo. If there's lead in the water, how they cook the freebie?
Sabrina	Don't you just need grease to cook fried food?
Pedro	You don't need water to cook freebies.
Joy	They don't need water.
Shakeem	Don't they gotta wash the damn stuff too?
Several	No.
Shakeem	We eat lead and we gonna get sick and that's gonna mess, that's gonna screw up, what's that one, the ex
Pedro	The excretory.
Shakeem	Yeah it's gonna screw up that one.
Derek	How?
Shakeem	You drink it and you get sick. Your bladder gonna get screwed up, now. You gonna get diarrhea.

In spite of some resistance to his idea, Shakeem persisted in raising this issue, using ways of speaking and interacting from his lifeworld outside of school. This transfer of practices between fields is agentic. His question was sound. If they wash the food and the water has lead in it, that will put lead on the food, which in turn will affect the body systems. The students had just learned the name and function of the excretory system a few days before, and Shakeem struggled to use that newly acquired knowledge and vocabulary. Pedro's assistance provides an example of shared effort in communicating. Illustrating communalism, the dialogue shows how students worked together to begin to construct an understanding of how a number of body systems (in this case the digestive and excretory systems) interact and respond.

Changing Participation

Among the *NSTA Standards for Science Teacher Preparation*, Standard 4 addresses the context of science and the need for teachers to establish connections between science and the personal, cultural, and social values of students. Standard 5 describes the use of prior conceptions and student interests to promote learning. The goal of relating science to the social context in which it is taught is laid out in Standard 7. These are not new ideas. What has been neglected is a critical discussion of what these goals might look like in inner-city high school classrooms with African American students. To assume that it will look like any other classroom, privileges white, mainstream ways of being and ignores and wastes the rich cultural funds that African American students bring with them. When inner-city African American students engage in science it may look unfamiliar, since they use cultural styles and ways unfamiliar to the teacher. Adherence to traditional practices shuts down potentially transformative practices and maintains the gate-keeping structure of schooling, which acts to keep most inner-city African American students from achieving success in school and on the job market.

In order to reduce this cultural bias, educators involved with teaching science to inner-city African American students must recognize, document, and publicize the science that these teenagers bring to the classroom. These students make frequent connections between their experiences and science, develop "ghetto analogies" for science concepts, and participate in science in ways that are significant but are often ignored or met with disciplinary action. These types of actions can be learning resources for the students, and they have the potential to be teaching resources as well. However, these practices of inner-city African American students have been largely overlooked. Inner-city classrooms are eventful places, where the students' contributions are often not heard or attended to by the teacher. At other times they are heard but are judged to be off-task, not meaningful, disrespectful, inappropriate, or just plain wrong. This is particularly true if the student is an African American male. In calling for an African-centered pedagogy, Murrell (2001) concurs that African American students' social, linguistic, and intellectual tools are frequently "misunderstood, unrecognized, and undervalued" (14).

Urban Underclass

Sociologists recognize the existence of an urban underclass and write about its emergence, consequences, and linkage to issues of race. Schools cannot ignore that for the first time most adults in many inner-city African American neighborhoods are not employed (Wilson, 1999b). Wilson contends that a neighborhood in which people are poor but employed is different in important ways from a neighborhood in which people are poor and jobless. He traces many phenomena common in inner-city neighborhoods, such as crime, addiction, family dissolution, and welfare, to the disappearance of work from these communities. The cumulative effects of their social and economic marginalization on the repertoire of dispositions and skills of the community members is profound. The current toolkits of this generation of inner-city, African American high school students are impacted by "the effects of living in segregated neighborhoods exposed to particular skills and styles of behavior that emerge from patterns of racial exclusion; the effects of attending lower quality, de facto segregated schools, and of being nurtured by parents whose own experiences and resources have also been shaped and limited by race" (Wilson, 1999a, 61). The structures associated with a poor neighborhood where few work impact the resources the community has and the ways in which its youth are equipped. The school system has failed to provide opportunities for a structure and set of practices to evolve that might counter this. Thus the cycle of social reproduction seems to be deepening. Schools do not routinely provide spaces where these students can either build new resources or learn to use the ones they have access to in ways that can open doors and make choices visible. When referring to young people in these marginalized, underclass communities, I believe it is a myth to refer to their options as "life choices" since there seems to be little opportunity for choice. And we wonder why they act "*that*" way.

Schools clearly play a role in gate-keeping, perhaps in science even more than in other areas, and I agree with Delpit that not to make the rules explicit (to the extent that we can) is to continue to play on a field that is far from level. However it does not seem possible to explicitly teach a large part of what this paper addresses, nor would we want the students to abandon their cultural toolkits. The rules of grammar can be made into lists to be hung on classroom walls, but the less tangible rules of participation, that are embodied in actions, cannot be so readily delineated. In addition the idea of explicit instruction focuses mainly on changing student behavior and neglects the need for transformation of the classroom system as well. It fails to recognize the dynamic dialectic that exists between structure and agency, and it is in this relationship that the possibility of change lies.

If participation in school and society is viewed as cultural enactment, it is dynamic, tacit, and instantiated in actions. Students can learn by participating in it. I believe that marginalized high school students must be allowed to participate in multiple ways—their own ways and in ways that are closer to mainstream. Such classrooms are possible when participation involves the use of students' own cultural resources as well as the acquisition of additional ways of participating over time. It is unlikely that this can be accomplished simply by exhorting teachers to "embrace diversity." Instead, teachers might turn possibility into capacity if teacher preparation involves cultural retooling and expanding; learning to "see" in new ways and enacting science curricula in a shared and negotiated way in inner-city science classrooms.

Educators and policymakers "rarely say anything about major rethinking of schooling or teaching practices . . . Few have examined how schools and teaching practices might need to change to enable students to succeed or how a closer look at student learning might trigger more effective responses" (Darling-Hammond and Falk, 1997, 190). We need to find new ways to teach that fit with what we know about urban schools and the students in them. One of the things we know is that inner-city African American students bring with them a rich cultural heritage and an array of experiences that can become a foundation of science learning.

Science teachers of inner-city African American students have tremendous opportunities to use the linguistic and metalinguistic facility, orality, and expressiveness of their students as foundations on which to build teaching and learning. Often these abilities are seen as impediments to be shut down or dismissed and the oppressive structures become seemingly concretized. Teachers react to micro-level gestures, tone, slang, and curses; this affects the meso-level unfolding of interactions and day to day events and translates to macro-level failures. Eliminating these patterns entails changing the ways teachers read and respond to their students, and not just to the Black English of African American teens, but to their ways of being. If we are to remove the gate-keeping of schools to enable underclass students to participate more freely in science and in society, we need to alter the nature of science teaching practices and science teacher preparation. We can begin with the language, models, analogies, and ways of expressing and representing scientific ideas that are seen as capital in our classrooms. Such norms are rarely negotiated and power is rarely shared in school. Rather, the imposition of norms is often considered a goal of learning (Yerrick, et al., 2003). Our adherence to the "same old" practices impacts African American teens who already come to school marginalized. If we continue to wonder why they act "*that*" way and expect them to act differently, we will continue to deepen their underclass position.

Chapter 7

Becoming an Urban Science Teacher: The First Three Years

Jennifer Beers

The first three years in the teaching profession can be a challenging odyssey of personal growth and professional development. In an urban school system, learning to teach is compounded by issues of identity, structure, and power, especially when the students are of a different race, class, and culture than the teacher. My professional narrative and the process of becoming an urban science teacher have taken me to two different high schools in the city of Philadelphia. My autobiography begins with my experiences as a student teacher, coteaching in a small learning community in City High School. This assignment laid the framework for my understanding of what it means to be an effective science teacher. Learning to teach in this school contradicted many of my past experiences as a student in high school and challenged my identity as a white, middle-class female. Teaching in this context also opened my eyes to larger systemic issues that serve to disenfranchise and marginalize the African American students I was teaching. It set the stage for my commitment to urban education and my desire to remain in the city after my student teaching.

As I moved from student teaching into a full-time teaching position, the context of my professional experience changed dramatically. I am now in my third year of teaching at a progressive urban charter school whose mission is to "give all students the skills they need to realize their dreams." This charter school is progressive because it is beginning to make important changes toward rethinking how we educate the youth of today. Both the administration and the teaching staff believe that the reform model being developed in the school will be viable across the city and, hopefully, across the nation. In this regard, these last two years have been immensely challenging both professionally and philosophically. Thus, in this new setting, I am not only continuing to learn how to teach science effectively, I am also part of a team that is making decisions toward what I believe is meaningful school reform.

Uncertain Beginnings

When I began as a student teacher at City High, I was not sure if I wanted to stay in an urban public high school or move to a suburban school district after my one-year teacher education program ended. In spite of my uncertainty, I had two major goals in mind. First, I wanted to help students develop their scientific literacy skills. I believe that the skills acquired through learning science and engaging in inquiry can help students reach their future goals. In a society that focuses on technology and scientific advancements, skills such as making observations, analyzing evidence, and constructing logical arguments can help students think critically about their world. I believe that if the goal of education is to help young people transform their lives and make change, then teachers must facilitate the development of the skills needed to meet this objective. My second goal involved nurturing students' natural curiosities and helping them make connections between science and their own lifeworlds. In emphasizing the relevance of science to the lives of my students, I hoped not only to make the discipline more interesting, I also wanted to create an environment where the students' ideas and experiences were the centerpiece of their science education. Though neither of these goals has changed over the last three years, my experiences of teaching in urban schools have been vastly different from anything I had expected. Finding ways in which I can meet my objectives has led me to examine my identity and prior experiences of learning science at both the high school and university levels.

   I am not sure that anything could have prepared me for what I felt or experienced as I walked into City High on the first day of school. I realized, right away, that I was extremely different, both economically and culturally, from my students. Moreover, I faced a harsh realization concerning the gross inequities of public education. Naively, I underestimated the systemic nature of the problems facing many urban public high schools. I began my teacher education with the typical rose-colored glasses of idealism and energetically began planning how to engage the students in knowing and doing science. I believed that good teaching and well-developed lesson plans with a variety of hands-on activities could involve the students in the kind of scientific discovery that permeated my own experiences as a student. At this point, I was not prepared for the types of difficulties that I would encounter, nor was I able to process my initial problems with engaging students beyond thinking that I was pedagogically unprepared.

In addition to what I was learning in the classroom, I was also regularly attending university classes, which provided me with the theoretical underpinnings to help me begin processing specific events and interactions. Though we were encouraged and even required to blend theory with practice, I found myself becoming frustrated at my inability to effectively process everything that was unfolding. How can I be expected to teach students about the relationship between velocity and acceleration if they are unable to do simple division? Who allows students to enter high school if they are still reading at a third grade level? How can we build this simple motor when we do not have working laboratory equipment? How can education be transformative when I can't even relate to my students' worlds? My list of questions seemed almost endless. Moreover, my search for answers seemed entirely futile. My first few months at City High were an emotional rollercoaster as I tried to process the cultural and structural contradictions that I was experiencing.

How Did I Get Here?

The impact of one's identity on the process of becoming a teacher cannot be understated. In thinking about this, I am reminded of Swidler's (1986) idea of the "cultural toolkit" or the cultural resources used to build strategies of action suited to particular situations or contexts. A large portion of my identity or "who I am" emerges from my prior experiences, my gender, and my cultural toolkit. Each of these large components impacts how I interact with individuals across a variety of social situations. They are the resources that I draw on in order to build my practice as a teacher and work together with my students. These components of my identity became sources of contradiction and difficulty as I moved between different contexts for teaching and learning during my yearlong teacher internship.

In Bourdieu's (1977) work on cultural and social reproduction, the appropriation of symbolic, cultural, and social capital can only be accomplished if an individual is familiar with the "code" used to decipher and utilize capital within a particular system of meaning. As a new teacher in an urban school, I was forced to move in and out of many different systems of meaning. Moreover, when I began at City High I did not have access to the system of signs and symbols needed to effectively interact with my students. Stark contrasts presented themselves as I moved between the events at City High and my experiences at the university and in the suburban neighborhood where I lived. My prior experiences as a student in an Ivy League institution with a diverse student body did not prepare me for the cultural and social interactions with students at City High.

> Today, I had to ask Mr. Jones to escort Thomas out of the classroom after a verbal blowup in the middle of class. I asked Thomas to remove his hood—a request that was immediately met with opposition and hostility. Thomas began to question why I asked him to do this. He kept saying "It ain't hurtin' nobody. Why do I gotta take it off? Don't you know it's part of my culture?" I told him, "It's the school's rules, Thomas, not mine. You need to take off your hood. You can't wear it in class." He retorted loudly that if I didn't care, then why should he have to remove it. I kept repeating that it was the school's rules. I swiftly began to see the futility in my argument; however, realizing we both made a spectacle of the situation, I knew I had to defuse it by asking him to leave. (Journal entry, October 19, 2000)

Thomas was a student in my class who I had great difficulty reaching both academically and personally. This exchange with him exemplifies a number of points that I needed to address with regard to my students. When the different cultural systems within the high school collide, this sometimes leads to overt signs of student resistance, as evidenced by Thomas's response to my request. His response also indicated his need to maintain respect among his peer group. In this situation, my inclination to invoke the school's rules simply exacerbated Thomas's unwillingness to take off his hood. It also totally disrupted the learning that was occurring in the classroom and highlighted my inexperience at classroom management. It was only after dialogues with other students and teachers that I came to understand that classroom management involved a delicate balance between being firm in my requests and gaining the respect of the class as a whole.

Playin the Game: A Lesson on Respect

Jen	So you know how to play the game?
Kareem	Yeah, yeah, basically I do. Like, I'm tryin to think of a good example. Like, I was, you got a principal right here, the principal Miss Wilson and you got a, let's see, a student right here and

you got a teacher right here. Out of all those three, the most respect I would give to is the principal. I would know how to talk to her. Like the principal, I wouldn't say, "Oh my God, man" and then turn my back. "See man this is why I don't like y'all, I don't like y'all teachers or y'all school," I wouldn't say that to the principal. I would say, uhm, "I'm having a problem with these teachers in here." I would get around it with that kind of respect. As a student, I'm talkin' to the student, I'd tell him to go blow something or go somewhere. I would give him not near as much respect as this person depending on the subject that me and the person were talkin' to.

Kareem's outward appearance—graffiti art on his jacket, tattoos on his arms, and a very large afro—would lead most teachers to believe that he is tough young man from the streets. His behavior contradicted this because he knew how to "play the game" that is associated with "surviving" in school. Through Kareem's working knowledge of appropriate participation in interactions with individuals from various levels of the school's hierarchy, he avoided disciplinary action, while also maintaining what is unique about his identity.

Kareem was a huge asset to me as I was beginning to develop my practice as a teacher. He challenged me to develop engaging science lessons like a whole class debate on cloning, but he also showed me the ropes when it came to dealing with the students in the classroom. His lessons on respect allowed me a window into his world and how some students were able to avoid the kind of trouble associated with not following the rules of the school. In many ways, what he was teaching me could be reflected onto my own practice and interactions with the students in our class.

Okay okay, here's what you got to do to with kids who act up in class. Are you listenin' to me? You got to give 'em that little pink slip of paper. Kick 'em out. Get rid of 'em. If you don't start using that pink piece of paper, they just goin' to walk all over you . . . You're too nice, Miss Beers. Too nice. And the kids can see right through you. You gotta learn to be cool but mean at the same time.

Kareem's advice was strikingly similar to what Delpit (1988) discusses concerning teachers' expectations of their African American students. In this regard, teachers need to maintain high expectations for appropriate behavior as well as achievement in the classroom. Prior to this conversation, I was reluctant to exhibit personal power of any kind since I was sensitive to building a rapport and relationship with my students. I believed that there was a contradiction between being strict, or using my authority, and creating a safe and nurturing classroom environment. Kareem's frankness about "playing the game" in school and his lessons on respect were messages to me that I needed to learn how to interact with students in a manner that was authoritative rather than authoritarian. Further, he also showed me that this kind of teacher behavior can lead to the establishment of meaningful interpersonal relationships which can foster student achievement.

One of the most important aspects associated with becoming/being a teacher is developing relationships, respect, and rapport with the students. Creating the kind of classroom environment that inspired collaboration and collective effort by all the stakeholders depended on my ability to develop social capital with the students. In any school environment, this can be a challenging task. However, in a classroom with students from vastly different cultural backgrounds and life histories, this task became even more difficult. In some respects, it required "cultural retooling," which would enable me to construct new strategies for action and interaction with the students (Swidler, 1986). My cultural repertoire and my preconceived notions about appropriate ways of interacting with individuals were no longer relevant. In some cases, I began to believe that my students perceived my identity through my clothes, my musical tastes, my knowledge of key sporting events, and my ability to decipher the messages encrypted in their slang.

> I was standing in the front of class today trying to begin the review for the test. We were going to play the game of Jeopardy to review the key concepts presented in the unit. The students had finished their question of the day and I was trying to get their attention so that we could begin the lesson. After my third attempt and an attempt made by Ryan, I decided to stop what I was doing and stand quietly in the front of class. Then, I heard "Dag, Miss B, is that a Polo you wearin'?" I looked down to notice that my skirt had the label on the outside. I quietly nodded and Derek began to ask, "Where did you get that?" Then he said, "Listen up everybody! Yo! Listen up! Look at Miss B, doesn't she look nice today. Look at her Polo. She's got style!" Derek's plea to the class and his attention to my clothing was just enough to get the class's attention and provided me with a window of opportunity. (Journal entry, November 6, 2000)

Throughout my time at City High, my experiences and interactions with my students pointed out that building social capital ultimately impacted how I was able to reach them in the classroom. At times, I felt constrained by my identity as a young white woman from suburban Philadelphia because I did not share the same cultural experiences as my students. Creating meaningful relationships with students on a personal level helped transform how I viewed myself and how I portrayed my identity to others both inside and outside the confines of the school. It also enabled me to make connections between the lifeworlds of the students and the science that they were doing in my classroom. In some sense, I was restocking my toolkit with different resources which I could use in my daily interactions. By the end, I did not feel as much of an outsider in the classroom and "who I am" was definitely different from when I started teaching at the high school.

Teaching the Biology Elective

It did not take me very long to realize that many of the students in City High had limited access to the types of science experiences characteristic of suburban or urban magnet high schools. For many of the students in the small learning community called SET, where I learned to teach, this was exacerbated by the location in the basement of the school and the fact that the science labs were located on the third floor. During my first semester at City High, I continually had to plan well in advance in order to gain permission to use a lab and the necessary science equipment. The structures for access to science resources limited my agency as a teacher to enact the kind of science curriculum needed to spark the interests of my students. In addition, my position as a student teacher sometimes exacerbated my difficulties with gathering resources for teaching.

When I traveled from the confines of the basement and my familiarity with the teaching staff in SET, I struggled to make liaisons with individuals who would help me provide resources for my students. The processes and "red-tape" that were in place for gaining access to science labs set up an explicit inequity in the kind of science education being afforded to my students as compared to other students located elsewhere in the school. In my mind, all science students should have equal access to labs and equipment so that at any given moment, students could test their ideas or hypotheses. I was appalled by the constraints that the system placed on doing spontaneous *real* science with the students.

As I began to confront these equity issues, a unique opportunity came up during the second semester at City High. Another student teacher, Ryan, and I were assigned to teach a biology elective course that would focus on human anatomy. This permitted us to plan an entire course and develop the curriculum to be used with a small class of fourteen students. It also provided an opportunity to work with Gale Seiler (see chapter 6), our methods instructor, on using a student-emergent curriculum that would stem from the students' lives and interests. Further, it gave us a chance to teach our science elective in a lab room, which was set aside for our science methods course. Interestingly, we found little overt resistance to our relocation because this science lab was hardly used. However, subtle forms of interference to our presence made me more aware of how my students were perceived as outsiders or interlopers on the third floor of City High.

Over the course of the science elective, the students became restless and resistant and, as a result, there were a lot of classroom management issues that were difficult to resolve. Resistance among the students was rooted in the manner in which they were rostered for the science elective. The students in the science elective did not have a choice in what class they were taking. Many of the students did not *want* to be in the class and they did not believe that it counted as part of their GPA. Therefore, kids would arrive late to class, argue with each other and generally reject our attempts at getting them to learn science. In addition, my own behavior in the class at times exacerbated discipline problems. As an outsider, I was wary of being on the third floor of the school and isolated from the NTAs and teachers that really knew our students. I became relentless in my attempts to stress the importance of appropriate behavior in the classroom and the need to stay in the room at all times.

As a teaching team, Ryan and I tirelessly attempted to find activities that would engage some of the students in order to counteract the inappropriate behavior exhibited in the classroom. Our plans for an emergent curriculum started out as a class discussion of what the students knew about the different systems of the human body and what they wanted to learn more about with regard to anatomy. Then, Ryan and I decided to start with the nervous system because it was the "command center" for the rest of the body. Ryan began with very traditional lessons on the neurons and brain structure using various hands-on activities, flow charts, and discussions to highlight the salient parts of the anatomy. Throughout this time, the students kept asking if they could dissect a human brain. Actually, they would have been content to dissect anything we could find. For over a month, their requests resonated in the back of my mind until, finally, Gale convinced me to order the necessary materials for dissecting a sheep brain. This decision made a significant difference in student behavior and served as a turning point for some of the students who were previously reluctant to participate appropriately in the class. The following journal entry describes how different things were when the students finally were able to carry out their first dissection (see chapter 6).

> The students really stepped up and took responsibility for what they were learning. I can't believe what a difference this made with regard to classroom management. There were no problems today and the students were on task for the entire period. They worked cooperatively and asked thought-provoking questions. Even some of the kids from Motivation SLC came into the class today.

They were curious to see what we were doing. NTAs and maintenance workers dropped in to look at what the kids were doing!! I really feel like we are making progress. And I don't think the students realize just how far they have come since the beginning of the semester. Drawing on student input is really paying off. These kids are doing the kinds of activities that science students SHOULD be doing in the classroom. I wanted to challenge them and they have accepted the challenge without question. I am really proud of what we did today. (Journal entry, March 13, 2001)

On several levels, the outcomes of this lesson exceeded my expectations for any activity that I had attempted during my year at City High. The level of student involvement and the kinds of questions raised by the students indicated to me that everyone in the class co-created an atmosphere of science, which promoted critical thinking and meaningful inquiry. It was exciting to see the students actively engaged; however, it was, perhaps, even more rewarding to see people from other parts of the school interested in what *my* students were doing in the class. Teaching the biology elective and enacting this lesson with the students was also a lesson for me on the importance of using their interests and aspects of their lifeworlds as windows into scientific discovery and learning. This is perhaps, one of the most important pieces of pedagogy that I carried with me into the next chapter of my new teaching career.

A New Context for Teaching Science

My initial uncertainty over where I would teach after student teaching was transformed into a commitment to stay in the city and teach in an urban high school. I wanted to be in a place where there was a great need for quality science education and where I could begin to address some of the inequities I observed during my year of apprenticeship. When I left City High I began teaching at an urban charter high school, Charter High, which is located in Center City, Philadelphia. This is a startup school with a mission to boost the achievement of urban high school students and provide them with a quality college preparatory experience. In comparison with City High, this is a vastly different context for teaching science, yet, it remained strikingly similar in the social, economic, and cultural backgrounds of the students. Yearly, students are drawn from the same pool of individuals as other schools in the school district. Thus, many of the students would have attended a neighborhood comprehensive high school such as City High if they had not enrolled in this charter school. In its first year, the school enrolled one hundred students and, now in the second year, the student body has reached two hundred individuals. This allows for smaller class sizes and more individualized instruction than is possible in neighborhood high schools such as City High. Further, the school's system of grading and an institutionalized culture of high expectations make Charter High different from most traditional high schools in the area.

A hallmark of the academic structure at Charter High is a nontraditional grading system in which students either achieve "Mastery" or "Incomplete" in a specific course. The science curriculum in the school is an eleven-course sequence in which students must master the content, as defined by state standards, as well as academic and scientific literacy skills. The curriculum is currently being planned so that each course spirals and builds on the previous courses in the sequence. Students are not permitted to progress through the sequence until they have gained mastery in all of the previous science courses. Thus, students are supposed to move at their own pace and gain a deep conceptual understanding of the scientific concepts and material presented in each course.

I was drawn to teaching at Charter High because I felt that the school would address many of the issues that I faced during my apprenticeship and that plague large, inner-city high schools. Teachers and staff would maintain high expectations for students and, hopefully, this would transform the students' expectations around what it means to achieve in school. By not allowing social promotion, the grading system alone might provide students the chance to understand the material and develop the skills for success. Throughout the school year, students are also given access to a number of different structures aimed at supporting them as they move through the curriculum in the school. For example, students are given the opportunity to take advantage of an academic support program that occurs after their core classes have ended each day. During the time allotted for the academic support program, students can meet with teachers to be tutored, catch up on missing assignments, or work on projects that were started in class. Near the end of each term, students at risk of not achieving mastery are identified and given a teacher-mentor who follows their progress through the remainder of the term. All of this has been put in place because there was an identifiable need to support students as they transition from traditional schooling to the more progressive nature of Charter High.

Throughout my first year at Charter High, the academic structure of the school began to respond to many of the systemic issues associated with urban public schooling. Many of the students who arrived at Charter High maintained low expectations for themselves with regard to the quality and quantity of work needed to achieve their goals. During the first months in the school, I struggled as I attempted to engage students in project-based group work that covered big scientific concepts and required students to apply what they were learning in my class. Some students would not or could not complete assignments that were different from the traditional worksheet-oriented activities they performed in middle school. Quickly, it became apparent that, as a school, we were butting up against years of low expectations faced by students in their previous schools. Moreover, no amount of social capital or rapport with my students seemed to help me in my endeavor; rather, it was a persistent and tiring effort on the part of all of the teachers to help students begin to develop high expectations of themselves and of their futures.

Teaching Each Other

A hallmark of the academic program at Charter High is an emphasis on a balance between project-based learning with authentic assessment and completion of standardized and other more traditional measures of academic performance. Thus, courses are structured around a final project, along with a final exam, which students must complete at mastery level. During my first year at the school, I was teaching a biology course on genetics and inheritance. This seemed like an excellent opportunity for students to begin learning about current issues in genetics research and how they are impacting society today. My intention for the final project was to have students develop a pamphlet that took a particular position regarding cloning research and exhibited their understanding of cell specialization and cellular reproduction.

At this time, the federal government was debating the utility of stem cell research and the ethical issues associated with human cloning. Many times, students asked questions about cloning, especially after reading about it in the local newspaper and hearing news reports on the television. Thus, this final project was a way to utilize student interest and draw on what they were experiencing outside of school. It was also an opportunity to guide students through the process of taking a position on a topic and using evidence to support this position. It raised the expectation level for what could be learned in a course designed to cover the basics of genetics and inheritance. I wanted the students to become more informed about what they were reading and hearing in the news. Finally, I wanted them to be able to decode the cumbersome scientific vocabulary and concepts discussed in forums associated with the scientific community at large.

The challenge for me was to bring to life the complicated scientific concepts associated with cell specialization and embryonic development in a way that was accessible for students. I did not want them to merely have a cursory understanding of the concepts; rather, I wanted them to be able to explain and evaluate the process of cloning and the outcomes of cell specialization in their own terms. As the project unfolded, I recognized that an obstacle for student learning would be the terminology associated with the process.

> I stood in front of the whiteboard and began drawing a matrix to help students better understand the difference between pluri-, multi-, and toti-potent stem cells. In the corner I placed the words "CELL SPECIALIZATION." I started by saying that where the cells start off in an embryo determines what they will ultimately become. I looked at the students in the class and they gazed back at me as if to say "What the heck is she goin' on about?" I then began to think, "Oh gosh . . . how am I going to explain this??" And then it hit me—basketball. How does a player like Allen Iverson get to be a point guard? Gale was sitting in the back of the classroom trying to help the students better understand the concept when I asked the question, "How did Allen Iverson become a point guard? Didn't he play football? Do you think he tried other positions when he was shooting hoops growing up?" One of the students in the class, Jack, understood where I was going with all of this and began talking about how A. I. had to specialize his talents to become a good point guard. He had to focus on certain aspects of his game like passing and the three point shot. I took that analogy and ran with it . . . eventually, students began to see how certain genes in a cell get turned on and off by virtue of their locations inside a developing embryo. It was truly an "AH-HA" moment for all of us. (Journal entry, March 10, 2002)

This was one of those teachable moments in which I cannot begin to explain how I was able to help the students relate cell specialization to someone becoming an NBA basketball player. In realizing that the analogy was working, the thoughts running through my mind were simply one step ahead of those of my students. Moreover, my students, especially Jack, used their knowledge of basketball to help me and the rest of the class co-construct a meaningful understanding of the process. In knowing my students' lives and cultural backgrounds, I was able to help them connect something familiar with something unfamiliar. The students walked out of that class talking about the different types of stem cells and, for the rest of the term as they worked on their projects, they helped each other clarify their understanding of the process. They often came back to the analogy as they worked on their pamphlets and, in many cases, would turn to their fellow classmates before coming to me with their questions.

I observed a lot of growth and change in my students as they progressed through their first year at Charter High. First, many became more serious about their work and began to assume greater responsibility for their own learning. They were excited about learning and going to school as evidenced by a very high attendance rate and incredible improvements in math and reading levels. Their level of questioning in science class and their willingness to help each other learn science was further evidence that the students were raising their expectations for their own learning. By the end of the year, the first-year students were excited about having a new student cohort and taking a leadership role in the classroom and the school as a whole.

While there was a great level of success in this first year, there were still several issues and questions left unresolved. The school lacked a science curriculum that extended beyond the first year. The scope and sequence of the science courses and the material taught still needed to be developed before we increased our student population. In addition, the curriculum lacked the "big picture" that would link each of the courses and create consistency with regard to the literacy skills being developed in each class. Further, there was a feeling among the administration and the teachers that the existing curriculum needed to be revised in order to make it more rigorous while also meeting the needs of our students. Although there were many students who achieved this year, there was still a group who had difficulties being academically successful. Thus, the challenge would be to systematically think about our students' abilities and create a course sequence that would effectively scaffold the skills needed to be successful in science and in school. As I moved in to my second year at Charter High, it was clear to me that our success as a school, as well as my practice as a teacher, would continue to be tested by new growing pains.

New Growing Pains

It is expected that the growth of a new school will bring with it fresh challenges and growing pains and the second year found teachers and administrators working to capitalize and build on our experiences of the first year. The focus of the second year was to provide new students with the structure they need to be successful in this different system of schooling. There is also a goal to document and standardize what we do as teachers to afford eventual replication of the school and its practices at other sites. Across the whole school that year, the science department team-taught an earth science curriculum. This curriculum was developed in a manner which would ensure that every student received a similar experience with science that is consistent with the competencies and standards developed by the school. In many cases, this created a tension between using teachable moments like those described above and keeping up with what my colleagues were teaching in their science classes. For me, there was an apparent pedagogical contradiction between the emergent science teaching I did at City High and teaching to this school's content and science literacy standards. These appear to challenge my agency as a teacher and the agency of students to use their experiences as a starting point for knowing and doing science in my classroom. In this regard, I am beginning to see that within any institution there are specific structures which might serve to limit my power to act as an educator. While I will ultimately be the individual who assists students in enacting the curriculum, I need to be mindful of the institution's goals overall. Thus, I continued teaching with some uneasiness about how the curriculum would be enacted and the kind of learning that would be occurring in my classroom.

The decision to teach earth science during the second year was generated as a result of a new standardized test required by the state. This has meant that both returning and new students were placed in the same class for science. Thus, the ability levels and expectations of the students varied along a huge continuum. In the first few months of the school year, I enlisted the help of the returning students to build a collaborative and cooperative classroom environment in which to learn and do science. In the first few months, I paired returning and new students in work groups for labs and classroom activities. In many cases, the returning students acted as tutors especially during lessons involving skills that they developed last year. This provided opportunities for students to co-construct their understanding of material presented in the class. In addition, I also found that returning students were quick to make connections between the big ideas in earth science and those we covered in biology.

> In a lesson on convection currents in the mantle, I was trying to explain that energy is cycled through the mantle, which then creates the mechanism for plate movement on the crust. Eric raised his hand and asked the following question: "Hey Miss B, isn't this like homeostasis, the earth's homeostasis? You know, the cycle is creating a balance in the heat." I couldn't believe the connection that he tried to make from last year! Although his analogy was a bit flawed, I asked him to continue by explaining what homeostasis was to the rest of the class, calling attention to the fact that the new students might not know about homeostasis.

I was truly amazed by how many of the returning students, like Eric, were beginning to make connections across scientific disciplines. They were enacting their prior knowledge to assist in learning new material in a new science course. For me, this was exciting because it was the first time that I experienced learning as an iterative process. I began to realize that drawing on these types of connections can put the students, as well as the science, at the center of their learning experience. Thus, there may be a place for students' interests and prior experiences amidst the standardized science curriculum at Charter High.

Final Thoughts

In the face of standardization and the push for teacher accountability, a student's deep conceptual understanding of science can only come from maintaining high expectations for achievement and providing them with an environment in which the teacher is learning with the student. For me, teaching is a lifelong journey of reflection, research, and change. Making adjustments in my pedagogy, rethinking my philosophy, and challenging my notions of "good science teaching" will forever be integral parts of my practice.

As I reflect back on my uncertain beginnings in urban science education and the struggles I am still facing with regard to my practice, I am certain of one thing—that drawing on students' lifeworlds and using student input can only enhance their learning. The most meaningful learning experiences for students in my classroom have often come from careful discussions and dialogues with them. Incorporating their ideas into classroom discussions and activities provides them with a sense of ownership of their learning. In this regard, I would encourage all new teachers to listen to their students and to be open to their suggestions because they are often experts on what it means to be a good teacher and can provide useful insights into what it takes for them to learn science.

Editors' Perspectives

Beers illustrates that effective teaching and learning of science in urban high schools takes considerably more than the teacher's knowledge of science. From the very beginning Beers experienced cultural and social otherness and had to learn to teach in ways that were adaptive to the social and cultural resources of her students. A key part of becoming effective was earning symbolic capital, notably respect, and earning the right to be considered the students' teacher. This involved Beers in interactions with human, material, and symbolic resources in which forms of capital were continuously exchanged in the production and reproduction of culture. A key part of Beers becoming an effective urban science teacher was her ability to participate in chains of successful interactions with her students and create and sustain environments in which their interactions with one another were successful.

When Beers focused on the control of her students she often met with resistance and when she built a curriculum around their interests, at City and Charter High schools, the quality of participation was much higher than when she enacted curricula based largely on her own planning and school and district standards. Getting students involved in coplanning of the curriculum seems to pay off and this is one of several chapters that suggest this to be a promising practice to use routinely. Seiler (chapter 6) and LaVan and Beers (chapter 8) provide further evidence of how Beers included the voices of students in the planning and enactment of science curricula. In chapter 9 Carambo illustrates the wisdom of providing all students with opportunities to pursue science that is of interest to them. Within a context of urban schools it appears to us that it is important for teachers to listen to students and show them respect by taking account of their suggestions. In so doing teachers offer students symbolic capital and increase the likelihood that they will be more willing to interact with the teacher and one another when what is being done is grounded in their suggestions and interests.

Chapter 8

The Role of Cogenerative Dialogue in Learning to Teach and Transforming Learning Environments

Sarah-Kate LaVan and Jennifer Beers

Current efforts to reform education emphasize the need for teachers to transform their educational goals, content, and instructional practices to make their curricula more appealing and welcoming to all students, particularly women and ethnic minorities (Bianchini, Cavazos, and Helms, 2000). Although many researchers and educators have answered these calls, highlighting the complexities involving teaching and learning to teach diverse populations effectively (e.g., Roth and Tobin, 2002), there continues to be a dearth of research to guide the evaluation of teaching practices and planning and enacting curricula in urban settings. This is particularly the case for schools in which the teachers and students differ ethnically, socially, economically, and culturally. Although these schools are often characterized as underachieving, based on achievement gaps that emerge from comparisons with suburban schools, such comparisons do not clearly connect with factors that might be changed with the purpose of increasing levels of participation and achievement.

Employing the lenses of cultural sociology and the sociology of emotions, this chapter explores the role of collaborative research that incorporates the uses of video analysis and cogenerative dialogues in creating a productive learning community in which an urban high school science teacher and her students transform their praxis. Since the use of video analysis, collaborative research, and learning to teach is a recursive-dialectical process, this chapter draws on two vignettes in order to highlight the theory and methodology and describe the outcomes that arose from this longitudinal program of participatory research.

The Need for Transformative Teaching

This research takes place over the course of one semester in Charter High, a small charter high school in Philadelphia. Currently in its second year of operation, the school enrolls approximately two hundred students in what is the equivalent of the ninth and tenth grade levels. The course in which this study took place was an introductory earth science 1 (ES1) class, and the teacher was Jennifer Beers, a coauthor of this chapter (also see Beers, chapter 7). One of the first courses in the science sequence, ES1 focused on providing students with foundational knowledge and skills necessary to move on to more advanced science courses. Since students came to the class with a wide range of home and school science experiences, they differed greatly in the capital and resources (human, material, and symbolic) that they could access to support their learning of science. For example, some students had experiences at home and in middle school that were a foundation for learning science, while others had experiences that did not so obviously support the learning of science.

When our research began, fourteen out of the fifteen students in this class were repeating the course because of their failure to reach mastery on a previous attempt. From a sociocultural perspective, ideological and physical structures found within the school, such as the nontraditional grading system, student resistance to being in school, home lives that impede the completion of homework assignments, and poor relationships with teachers, were possible explanations for the students' academic struggles. For these reasons among many others, a majority of the students not only resented the fact that they were enrolled in this course again, but also felt that, since they had been exposed to the course content, just showing up for class every day would allow them to master the material needed to move on to the next course in the sequence.

Compounding the problems of students having negative emotions about the class were difficulties associated with the history that some students brought with them regarding their teacher and her teaching practices. In the past, many students had come to know her as an "interfering, authoritarian teacher," who did not value the culture that the students brought to the classroom. Additionally, the students perceived Beers as a teacher who did not understand that their experiences outside of science class and school impinged on their learning while in her classroom. Throughout our research, the students often pointed out that the porous nature of the classroom field allowed culture from external fields to enter, yet when they enacted practices and dispositions (Boykin, 1986) from external fields, Beers "shut down" these practices in order to settle students and regain control of the class. These feelings are illustrated in the following excerpt from a conversation of a student researcher with Sarah-Kate LaVan.

Ace   Ms. Beers was mean. Nobody really liked her.
LaVan   Why not? What did she do?
Ace   A lot of things. She would just yell at us or bark orders at us. She would tell us that we needed to sit down and listen to her. She didn't let us ask lots of questions or challenge her, like her ideas or what we were doing or that we didn't really get it. And she didn't get that some of us had stuff outside of the science to do so we couldn't do every homework that she gave. But when we tried to explain it to her she just told us to get it done. When we started doing cogenerative dialogues, she started to listen and figure out what we could all do to get our work done.

Beers was unaware that by using "shut down" practices, such as the ones described above, she often caused symbolic violence to many students. This resulted in animosity and frustration toward Beers by the students and anger from Beers toward the class as a whole. Thus, many students ultimately resisted, and Beers lost the social and symbolic capital necessary to teach effectively.

The structures of the school field presented contradictions for Beers as she endeavored to reflect on possible pedagogical strategies that would build from what she could do, connect to the cultural capital of students and be compatible with the rule structure that was set in place by the school. Beers had to account for a range of student educational and personal experiences, consider how to engage students who had negative emotional energy associated with her teaching, and meet the needs of students who had already taken the course at least once. Given this background, this study addresses the following questions:

- In what ways do collaborative research and the use of videotape and associated digital resources enable a teacher to transform her praxis?
- How can this type of research allow all stakeholders to take an active role in creating classroom structures that are more conducive to meeting individual and collective goals?

- What role can collaborative research play in the current arena of educational reform?

Capturing Activity in the Classroom

Using Collaborative Research to Access Data Resources

This critical ethnography utilizes collaborative research and video analysis (i.e., using videotape of classroom activities and video editing software) to identify patterns of coherence and contradictions in classroom activities, guide the direction of the research, lend the perspectives of the participants, and transform the field being studied. The interpretive nature of the research allowed for the focus of the study to reflect the voices of all participants and activities within the classroom. It challenged us to change power relationships and the division of labor not only within the research but within the classroom as well. In order to do this, a research team comprised of LaVan (a university researcher), Beers (the classroom teacher), four student researchers (Ace, Terrell, Shania, and Derek) and ten student participants was initially created to inform our research and class activities by developing and analyzing data resources that included videotape of classroom activities, field notes, journal entries, interviews, and cogenerative dialogues.

Cultural sociology and hermeneutic phenomenology (Ricoeur, 1991) informed the selection of student researchers and various other participants in the research. Since the class in which this research took place was one of the first science courses taken by students at Charter High, many of them enacted cultural practices (habitual and intentional) and schema (beliefs, values, and ideas) that Beers did not value or consider valuable to classroom learning. As a result, many students were prevented from participating in the class as they would prefer and became frustrated in their endeavors to learn science. Other students seemed to adjust and managed to align their practices with Beers's expectations. Accordingly, in an attempt to bring varying voices and perspectives into the research it was imperative for us to choose students from a wide range of cultural, social, economic, and science backgrounds.

In order for us to accomplish the selection of student researchers we adopted a hermeneutic-dialectical process. For example, since educational research in the United States illustrates that low socioeconomic status is associated with low-level experiences with science resources (human, material, and symbolic) we selected two students from differing economic and cultural backgrounds, who also differed in gender and their orientation toward academic achievement. As the study progressed and more students from the class became interested in the research and changing the classroom structures, additional students became student researchers and/or participated in cogenerative dialogues in which we used video resources to examine and analyze classroom activities. By the end of the study, all but two students regularly participated in cogenerative dialogues, examined videotape of classroom activity, and openly reflected on practices used by Beers and the students.

Using Video to Capture Classroom Activity

A growing number of researchers and teachers in science education have begun to understand the benefits of capturing and evaluating classroom activity through the use of videotape (e.g., Roth and Tobin, 2001). Among the many advantages of using videotape as a data resource is the manner in which it allows researchers and teachers to review activities that are often missed due to the unfolding nature of social life and participants being unaware of practices that are habitual. By replaying videotape, participants are able to discern patterns in practices and associated contradictions that occur in the classroom. In this context, videotape is a powerful tool in that it provides a springboard for discussion during cogenerative dialogues and serves as a source of reference for changing classroom activities and structures.

During the course of our research, classes were generally recorded two to three times a week (an hour per class). However, when necessary, taping increased to four or five times a week, as was the case when students were working on projects, laboratory activities, or when Beers needed another adult present in the classroom. Classes were recorded from varying positions within the classroom, and due to the collaborative nature of the research, which required LaVan to coteach, students often took charge of videotaping daily activities. On these occasions, the focus of the videotape depended on the interests and practices of the student. Occasionally, there was a need for all stakeholders to be involved in the classroom activities, and therefore the camera was placed either in the back or front corner of the room on a tripod and left to record.

Through the use of the video editing program iMovie3, we are were able to identify practices and contradictions on a meso level and analyze interactions at a micro level, where we could slow the video to tenths or hundredths of a second and break down the practices and interactions as they were enacted. In this way, we were able to focus on verbal interactions, gestures, bodily movements, physical spacing, and orientation of participants, and the emergence of solidarity and mutual focus among members of the class. Co-relating microscopic- and mesoscopic-level analyses allowed us to further understand how these individual and collective practices and interactions created structures that afforded teaching and learning in this science classroom.

Cogenerative Dialogue and Classroom Activities

In this study we assume that each participant brings unique understandings and experiences to a field of activity while experiencing and interacting with structures in different ways (Ricoeur, 1991). For these reasons, when we were evaluating classroom activity and attempting to restructure learning environments, it was crucial not only to consider the perspectives of all stakeholders and examine the relationships between the individual and the collective, but also to understand that differing perspectives could not be brought into one overall understanding or experience about the classroom activities or structures. Cogenerative dialogue (Eldon and Levin, 1991) allowed us to retain diverse perspectives, value difference, and promote solidarity and cohesiveness around shared schema and practices.

Evolved from the coteaching studies of Roth and Tobin (2002), our approach to cogenerative dialogue used video vignettes selected from recordings of classroom activity to promote the emergence of cogenerated understandings and collective responsibilities for agreed-upon decisions about roles and insights into possible ways to distribute power and accountability in the classroom. Video clips were selected from these tapes by various members of the research group, on the basis of their salience to the quality of teaching and learning in the class. For example, when student researchers selected, decisions about salience were made in accordance with their understandings of social theory (which we taught them) and included such constructs as agency, structure, resistance, and social and cultural capital. Videotaped incidents were also selected if they particularly captured the group's attention and interest.

We regard this type of cogenerative dialogue as a catalyst for improving the quality of the learning environment because it creates structures in which participants have voice in selecting video vignettes, articulating and explaining personal experiences, as well as sharing responsibility for co-constructing understandings and orientations toward action about these shared events and classroom structures. The power of this type of dialogue also lies in the opportunity for participants to identify and review practices that are unintended and habitual, while discussing the power relationships, roles, and agency of all of the stakeholders. The associated redistribution of power (vertically and horizontally) allows all stakeholders to discuss future actions and activities as well as aid in planning for improvements to the quality of teaching and learning.

Cogenerative Dialogues in Small Groups

Cogenerative dialogues can take many forms and are employed in a variety of circumstances. Since we wanted to establish an informal, sharing atmosphere in which to review videotapes and discuss classroom activities, our cogenerative dialogues were generally held two to three times a week during the students' lunch period. Usually lasting about half an hour each, there was no standard manner in which we began the meetings. To help develop power relations and a division of labor that was more equitable, rules were created to support and encourage all participants to raise issues and analyze videotape. Generally, during each meeting, we reviewed videotape, identified salient clips, discussed issues that were of concern to the group, or examined specific incidents that occurred during class time.

During these meetings, students did not just talk about classroom activities, but spoke about many other issues such as relationships with peers and other teachers, and school structures that they felt prevented them from succeeding. The small group discussions with student researchers as well as larger whole class dialogues were structured so that all participants were free to speak openly with the goal of assuming collective responsibility for the quality of teaching and learning in the classroom. These discussions helped to minimize the power differences between students and coteachers and assisted us with finding solutions regarding what was to happen in subsequent science lessons. The variety of topics we discussed also allowed us to see how students placed the activity of learning science within other activities and to enrich our understandings of the practices that other teachers and students employed.

The following description and transcription of a cogenerative dialogue is presented as a means of understanding the potential of using videotape and cogenerative dialogue to catalyze changes in classroom practices. On this particular day, LaVan and Beers chose the topic for this cogenerative dialogue because two factors alerted us that classroom events were not progressing as well as they might. First, talking before the meeting, Beers recounted an incident that occurred while she was absent from school the previous day in which students had disrespected a student teacher. During this incident, the students exhibited resistance to being in the classroom with the teacher, which resulted in one student's removal from class and the others not finishing their assigned work. Second, upon reviewing a videotape of classroom activity from the previous week, La-Van noticed pervasive negative emotional energy and a lack of solidarity that was evident on meso and micro levels (noted by eye rolls of students, asynchronous head nods and gestures, and an absence of entrainment between the students and Beers).

The discussion began with LaVan explaining about the negative emotional energy she had observed in the videotape of classroom activity. Using video vignettes students had selected during previous cogenerative dialogues as examples to illustrate her points, LaVan then raised the question of whether the students saw examples of resistance or negative emotional energy. Building on LaVan's ideas, Beers provided concrete examples of resistance, citing interactions between some students and the student teacher from the day on which she was absent. After the students recounted what had transpired, one of the student researchers (Ace) spoke up. Taking responsibility for his actions, Ace noted that some students, including him, "purposely disrespected the student teacher and acted out." In explaining his position, Ace began with a detailed description of his actions and those of the teacher. Then, carefully supporting his assertions with concrete examples of practices enacted by him and the teacher, Ace blended observations, descriptions, and feelings about why he had chosen to be disrespectful.

Valuing cultural sociology and its usefulness in examining issues such as these, Beers and LaVan spoke to the group about the student teacher's role in this incident. In doing so, we highlighted the belief that structures created by participants are dialectically related to and afford the agency and practices of others. Therefore, instead of focusing on the negative emotional energy produced in the classroom and placing blame on the student, the discussion moved from a description of what had happened to an analysis of the practices that could be enacted in the future to prevent Ace from getting kicked out of the classroom and others from being seen as "acting out." Feeling strongly about this topic, Terrell focused the discussion in a new direction and brought up the practice of code switching.

Terrell	You do it.
Ace	(*Highly charged negative emotional energy, loud voice*) =I guess I ain't gonna be nuttin' in life if I can't do that.
Terrell	(*Loud voice, body positioned toward Ace*) =But you do it and you don't know you do it. [You don't act the same
Shania	[But don't have a negative attitude about it like, look at it as something positive like
Terrell	[You don't act the same. No go ahead. (*Nodding to Shania*)

Shania	Like when somebody realizes that they bad then they change their behavior . . .
Ace	=I ain't gonna be nuthin' in life then.
Terrell	Do you go to church?
Ace	No.
Terrell	You don't go to church?
Ace	Yeah, I go to church sometimes.
Terrell	=Well, you don't act that way when you go to church do you? You code switch. You don't realize you do it, you just do.
Ace	Yeah, but this is different. (*Voice gets louder and more charged with emotion*) I'm not gonna give up who I am for her. I won't sell out.
Terrell	You don't have to (0.5) you just gotta code switch to get by.
Beers	(1.0) Think of it as playin the game.

It is in this excerpt, and specifically the talking through of future strategies, that we find one of most significant arguments for the use of video analysis and cogenerative dialogue. We see the students gaining and developing skills associated with scientific argument and fluency. Focused on one topic at hand, the students clearly have solidarity and the common goal of getting Ace to understand that he possesses capital that will allow him to succeed in the science classroom as well as in the world. This is noted on a micro level in the videotape, by synchronous head nods, eye gaze, forward-leaning body positions, and mutual focus and, in the transcription, by the single topic of discussion, overlapping speech (by Shania and Terrell), and building on each other's ideas. Examining the groups' interactions on a microscopic level illustrates that the student researchers value this activity as an important process for both the individual about whom they are speaking as well as the collective.

Equal turn taking, active listening, and sharing of responsibility among participants in constructing their discussion also illustrate the re-distribution of power and agency that the group feels is necessary for the individual and the collective to be effective in attaining their goals. An example of this is illustrated through the interactions of Terrell, Shania, and Ace. Realizing that they are speaking over each other and they would not be heard, Terrell defers to Shania and tells her to "go ahead." It is only when Shania is finished making her argument that Terrell begins to speak again. Recognizing (by Ace's rebuttal) that he had not listened to what Terrell and Shania had said, Terrell changes his tactic and this time supports his argument with an example of when he believes Ace would have needed to code switch in his everyday life. Noted in the transcription, the students not only take responsibility for educating Ace and begin their discussion without prompting from either Beers or LaVan but they also control the direction of the discussion through equitable turn-taking, declarative statements and assertions as well as by carefully listening and building on one another's arguments. This is most evident when Terrell changes his argument strategy in favor of another. Additionally, as illustrated in Terrell's arguments of when Ace code switches, the students often use practices which draw on the cultural capital and experiences of their peers to have them understand other's perspectives. These are all practices that put the students on a pathway to developing science fluency as well as for becoming productive members of society.

Using vignettes selected from videotape not only served as a visual aid to focus the group's attention, but also as a referent for individuals to build shared language and understandings about classroom activity and issues. Although not directly illustrated in the vignette above, it is not uncommon for the group to refer back to specific instances in video vignettes throughout cogenerative dialogues in order to gain others' attention or understanding. In this vignette, the mere mention of resistance and negative emotional energy (noted as asynchronous eye gaze, head turns, attention to speakers, etc.) were adequate resources for the group to begin to draw on their prior understandings as well as shared language and experiences. This points to the importance of video in identifying and discussing the symbolic understandings participants have and their salience to classroom activity. In addition, video vignettes provide a context for students to share their culture and explain how practices and possibly schema are useful to them in affording agency. That is, during cogenerative dialogues, video vignettes allow students to identify cultural resources and argue for their use in the classroom, as forms of capital that have potential relevance to their learning of science.

On a final note, the intimate, relaxed nature of small group cogenerative dialogues generally provided us with time to get to know the students personally, develop social and symbolic capital, discuss topics that were pertinent to the lives of all stakeholders, and come to understand the various perspectives that each person in the group represented. Although our discussion did not lead each participant to the same final resolution, this incident in particular afforded all of us the opportunity to discuss and come to understand the significance of respect as a topic that was central to the students' lives both in and out of the classroom. As a result, we were all able to come away with the understanding that respect and disrespect are complex concepts and that people often make errors not because they want to, but because they are unaware of the culture of the other.

Whole Class Cogenerative Discussions

Small group cogenerative dialogues, which took place during students' lunch periods, were videotaped so that verbal and nonverbal interactions could be analyzed in an effort to create an effective forum for transforming the curriculum to better meet the students' needs. Often, the video from these smaller cogenerative dialogues or the clips from classroom activity served as artifacts for whole-class cogenerative discussions. Depending on the issues at hand and the student researchers' levels of comfort, topics were either raised in the small lunch group or video clips were chosen to be discussed with larger groups of students during whole class dialogue. Although at the beginning of the study the research group had expectations of bringing video clips to the whole class discussions once every week and a half, school structures placed time constraints on classroom structures and restricted the amount of videotape we could show during class time. Therefore, the use of video vignettes during typical classroom instruction time was limited to one period every four to five weeks.

The vignette and the transcription below represent an example of a whole-class cogenerative dialogue that transpired as a result of reviewing video within a small lunch cogenerative activity. Upon examining video of a whole-class interactive lecture and discussion, we noticed that many students were exhibiting negative emotional energy and were not paying attention to each other or to Beers. Believing that gaining solidarity and collective responsibility were two crucial aspects for improving classroom activities and learning, we decided to share the videotape with the rest of the class and use student practices as a point of focus from which to begin our discussion.

While the students and Beers were seated at tables preparing to watch a videotape, LaVan (standing next to the television at the front of the room) explained that sometimes, in order to pick out patterns and gain a better understanding of what is occurring in the classroom, previewing the tape through fast forward is very useful. Then, playing the videotape on fast forward LaVan drew the class' attention to the students' actions, particularly who was participating and resisting.

Class	hhhh
Derek	Ah man.
Chantal	He is doin' everything but listening.
Class	hhhh
	(*Derek looks back at Shania. She raises her hand and smiles, as if to say that she has no idea why she is acting the way she is on the tape.*)
Chantal	Donald, you the worst. Look at me. I'm combin' my hair and everything.

Class	hhhhh (loudly)
Chantal	But I'm still looking forward though. (hhhh)
LaVan	So Chantal is there combin' her hair and Derek just disappears.
Class	hhhh (still focused on the television) (2.3)
LaVan	How many people do you think there are paying attention and taking notes?
Shania	I was paying attention.
Chantal	I was [taking notes after.
Shania	[I was paying attention. Look, I raised my hand and was taking notes the whole time.
Derek	=I do it in my own little way. See there.

After this segment, the students laughed for a few seconds and then suddenly became silent again. There was clear evidence of the class' mutual focus and interest in watching the videotape as illustrated on the micro level by the synchronous smiles, laughing, head nods, and eye gaze toward the television. The relaxed, forward-leaning body positions of most participants and the comments about students by students not only suggest a comfortable atmosphere, but also a sense of solidarity and the community's acceptance of its goal of critiquing classroom activity to improve the quality of teaching and learning.

As directed by a student, the comments quickly shift from the students' actions to Beers's practices in the classroom. As can be seen below, the frank nature of student observations and their critique of Beers's teaching practices, in addition to student-guided shifts in topic, further illustrate the relaxed nature of this activity, student agency, and the redistribution of power within this activity.

Chantal	Look at Ms. Beers.
Class	hhhh
LaVan	What does she do that's kind of cool or interesting?
Chantal	She paces
Beers	=I pace a lot
LaVan	Does that make it interesting at all or no?
Terrell	=It look like she's nervous.
Chantal	=I don't know, you can stay focused but you can't stay focused because you gotta keep makin' sure that you don't [miss
Beers	[hhhhh. You certainly can't fall asleep in my class.
Class	(still watching the video) hhhhh
LaVan	(hhhh) Maybe, I don't know. Some people seem to be doing it pretty well.
Beers	(4.0) Would you rather me stand in one place?
Class	hhhh
Chantal	No
Terrell	No
Beers	No? (2.0) Okay.

After some time and further conversation about student practices, the focus returns to Beers. Pointing to the television, Chantal draws the attention of the class to Beers. She builds on comments from one of the previous topics and makes the assertion that Beers not only bounces around the room, but she also consistently walks to the back making sure to check on students, who due to their position in the class would otherwise be neglected.

Chantal	Ms. Beers is the only person that I know that walk in the <u>back</u> of the room.
LaVan	(0.3) Is that bad?
Derek	=See look at that. (*pointing to the screen*)
Chantal	=You look at <u>that</u> . . .
Beers	Are you saying that teachers normally stand in the front? (*students' heads turn toward Beers*) Is that bad to just stand there? (*meaning the front of the room*)
Chantal	=Yeah, and I be sittin' there like (*puts her head on her hand and leans toward the table like she is tired*) [And after a while it starts soundin' like (*imitating a slow, deep voice*) yooooou doooo myyy
Ace	[Yeah, huh Ms. Beers, she all over the <u>place</u>.
Beers	Yeah, I've ants in my pants for fear of that. [I don't like standing still.
LaVan	So is it a good thing?
Beers	Do you like it when I walk around or don't you?
Troy	=Yeah. I like it because it's hard for me [to fall asleep
Shania	[but it's hard to fall asleep because you bounce (*pointing from place to place in the classroom*) all around the room.

Since issues raised in whole class cogenerative dialogues are usually decided by consensus within the research group, while watching videotape of classroom activity or through discussion of classroom activity, the topics brought up are the ones that are pertinent to the success of the students and Beers. As a result, all participants generally believe these issues to be of concern and valuable to the health of the classroom community. As evidenced by mutual focus of attention (synchronous eye focus, head turns toward the speaker, and laughing), as well as active listening to ideas and critical feedback regarding practices, solidarity and positive emotional energy are typically reached through discussion. The videotape serves as the source of discussion, a potential mutual focus, since it affords all participants with a common referent to deconstruct shared events and the practices of both the individuals and the collective. As depicted in the vignette, the use of video provided participants with a structure in which they could make observations about their learning practices and Beers's teaching practices, as well as find evidence to support their assertions.

who's disrespecting, who's resisting, who's acting like an ass. We can see what Ms. Beers is doing and help her change to teach us better. Make class better.

As expressed above, Terrell understands the benefits of reviewing videotape as one way to improve the agency and success of the collective, without which teaching and learning would not be as productive.

Watching videotape and holding cogenerative dialogue were generally valued times for the students. We often had requests for additional cogenerative dialogues to be held (during lunch and in the classroom) in which to review videotape or discuss issues of importance to them. One student researcher, Ace, eloquently pointed out that these discussions not only benefited the classroom activity and the collective but also the individual student in that it gave him or her a voice and

> a chance to hang out with and get to know teachers and students in a very real way. We get to build social capital and have fun at the same time. We are not sitting there being fake, but talking about interesting things that relate to all of us.

By watching video, students were able to develop skills associated with science fluency while discussing issues that were important to them. Discussions generally encouraged students to use multiple and varying resources (human, symbolic, and material) while observing, describing, and communicating understandings of the circumstances to others, supporting ideas with evidence, and listening to others' perspectives. This was most evident in the vignette of the whole class cogenerative dialogue as students examined practices enacted by both Beers and the students. Students used the videotape as a referent to observe, make assertions about, and draw examples of practices of all participants. Students also recognized that they were developing new practices and emerged from these discussions with a greater understanding of teaching, learning, and the world around them.

Editors' Perspectives

The realization that cogenerative dialogue is a field that is structured and associated with cultural production, reproduction, and transformation provides exciting insights into its potential applications. As we enacted cogenerative dialogues initially we focused on small groups in which representatives from each stakeholder group participated in a conversation in which no voice was privileged. In selecting participants we focused on the inclusion of diverse voices so that the learning potential was maximized. As our applications of cogenerative dialogue expanded it became apparent that collective responsibility for processes and outcomes were central, hence a priority was the focus on cogenerating consensus on outcomes and responsibility for enacting them.

The enactment of collective decisions occurs in another field (e.g., a classroom) and the idea that students and teacher (and other participants in the cogenerative dialogue) had a shared responsibility for successful enactment was radical and in many ways counter to traditional views of classroom practice. No longer did it make sense to hold the teacher accountable for the quality of teaching, or the students for the quality of learning. Joint responsibility drew attention to the significance of the different conditions that support a teacher and students interacting in ways to produce solidarity as a means to afford the emergence of science fluency throughout a classroom community.

It has been customary in education reform to focus on what teachers need to learn of the social and cultural capital of students. Cogenerative dialogues obviously are effective sites for this to occur. In addition, cogenerative dialogues are sites for students to learn of and experience the social and cultural capital of their teacher. In so doing they can build adaptive practices that increase the likelihood of successful interactions across the social and cultural boundaries such as those that characterize the classroom field. That is, cogenerative dialogues are sites for students and teachers to communicate and otherwise interact across the boundaries of age, ethnicity, class, and at times gender. In so doing participants learn to create successful interactions across such boundaries and, having done so, there is a potential for them to enact similar practices in the field of science education. Accordingly, cogenerative dialogues are sites for the generation of capital that can potentially be enacted in science education, thereby providing a structure for others, who were not participants in the cogenerative dialogue, to learn by coparticipating alongside those who were involved. We regard cogenerative dialogues as activities with high revolutionary potential.

Chapter 9

Learning Science and the Centrality of Student Participation

Cristobal Carambo

I Don't Have It All Figured Out Yet

It is early in October of the school year. I am watching videotape recordings of my chemistry class. We are discussing the science content and procedures for a lab called "Solubility and polarity of compounds." The lab activity is structured so that students first observe the solubility of salt and naphthalene in water, turpentine, oil, and alcohol and then use electronegativity data to determine the type of bonding that characterizes each of the solutes and solvents. It is my hope that they will discover the connection between the nature of the intermolecular forces and the observation that the ionic compounds dissolve in the polar substances but not in the nonpolar turpentine. The laboratory will end with a challenge in which students explain the reaction of food coloring (a polar substance) when placed into mixtures of polar and nonpolar solvents. As I continue with the introductory mini-lecture, I am pleased to see that the class is easily managed. Although these are the same students that proved so difficult last year (my first year at City High School), my capital in the community has grown to the point where I have few behavior or management problems. Whenever a student speaks out of turn or engages in inappropriate behavior, a slight look or a word is enough to stop him or her.

Once instructions are over, we set out to do the lab. The students arrange themselves around six large worktables that serve as lab stations. I have set the materials out on the preparations table and I assist as each group sends the "materials person" to the table to get equipment for the lab. Although we have discussed the lab, and each table has several copies of the instructions on it, I notice that the groups fail to begin the lab. The procedures seem to confuse them and they cannot figure out what to do. They sit idly at their workstations, "joke about," toy with the pipettes, and call out to me for help. I am called from table to table for instructions as students complain, "I don't get this," "This is confusing," or "What do we do next?" Some attempt to work, but they mix substances without paying much attention to the lab procedures. When I approach a group, the situation improves somewhat; I review the concepts and procedures, organize the division of labor and the materials, and encourage students to continue. They then begin to work, however, a few moments after I leave, the group falls into listlessness once again. The class continues in this fashion for the remainder of the period. I continue to move from group to group encouraging and explaining. However, no student is able to fully make the connection between his or her observations and the type of chemical bonding, nor can anyone explain why the food coloring settled in the polar layer of the mixtures. Although the students behave relatively well, they do not invest the kind of intellectual and scientific endeavor needed to construct important concepts on their own. The quality of the work in this lab is such that there is little possibility of using the activity as a basis for understanding the behavior of polar and nonpolar substances. I am perplexed by the lackluster quality of their work because I know that these students are resourceful, intelligent, and energetic. Why then is their work devoid of personal energy? Given the rapport that existed between us, I wondered how I could foster the kind of learning environment that would encourage students to fully engage in their learning. The remainder of the year was devoted to improving the quality of student participation and learning. This chapter is an account of the theoretical perspectives that informed the changes that were made to my science curriculum and the associated changes that emerged in the practice of science in my classroom.

Changing the Classroom Structure

I am in my second year of teaching science at City High. Now an accepted member of the community, it has been some time since I have had serious disciplinary or classroom management issues. Given the positive nature of my relationship to these students, I had assumed that I would be able to foster the kind of learning environment where students would be active participants in the creation of their knowledge. However, events in my chemistry class this year showed that this would not be so easy.

As I watched videotapes of my class sessions early in the year, I realized that most of the students wanted to participate, but some structural element was impeding their ability to do so. Research that had been recently done at City High suggested that students responded positively when their suggestions were used in the creation of science curriculum (Seiler, 2001; Seiler, Tobin, and Soko-lic, 2001) or when the science content related to pertinent issues in their lives (Barton, 2001). With this in mind, I asked my chemistry students to suggest top-ics or questions that they might wish to explore in their chemistry class. We used the lists to cooperatively design a series of activities and laboratory investiga-tions that we would carry out as part of our chemistry curriculum. The nature of some of the labs presented a problem as they had an organic focus and our school had neither the chemicals nor the facilities to perform organic chemistry. We were fortunate to have access to a local university's chemistry facilities and a student teacher as part of our class. We selected a small group from the class to accompany the student teacher (Mr. Chris) to the university twice a week to per-form some of the requested experiments. The laboratory activities carried out at the university included:

Synthesis of aspirin
Acid base titration
Making soap
Christmas candy
Separation of the ingredients in pain relievers using chromatography
Extraction of caffeine

The remaining questions and topics were covered as part of our regular class time.

Videotapes of the labs done on the university campus showed improvements in the students' attention to laboratory procedures and the use of equipment as well as an increased ability to personalize the science content and complete labo-ratories on their own. An example can be seen in a short vignette of the labora-tory on the synthesis of soap. The video shows one student making jokes on camera. He purports to be a "gangsta chemist" who is working with deadly chemicals for heinous purposes. He "toys" with the deadly sodium hydroxide, feigns an explosion while attempting to light the Bunsen burner, and takes the camera person on a tour of the secret materials located in the laboratory. Al-though it seems that these behaviors may have detracted from his ability to learn from the lab, the student was able to successfully complete the procedure for the synthesis of soap during the allotted time period. He also helped other students with the procedure, mastered the use of the flint striker and the Bunsen burner before anyone else, and engaged the student teacher in a discussion over the best method to increase the yield of the product. In watching the videotape, I noticed that all of the students had a grasp of the laboratory procedure and purpose. There were relatively few procedural questions, as they did not seem confused about the equipment or materials. They showed genuine curiosity about the chemicals and their work. Although there were many jokes and side conversations, there were no behavioral or management problems and students were on task for the entire laboratory. What struck me as most interesting was that they seemed to genuinely enjoy themselves. The videotapes of two other laboratories revealed similar patterns in terms of students' levels of engagement and personal energy. Students reported that the labs were "fun," and that they got to learn in interesting ways.

Mr. Chris	So what did you learn?
Randy	We learned but we learned but we learned to play with things in a certain way. We learned to play, that's another way of learning.
Mr. Chris	So what did you think of the antacid lab?
James	It was dry, it was boring . . . No actually I did like this one, the next day when we got back to school and we were able to figure out which one was cheaper, so we could actually know which one worked best and was worth the money.
Mr. Chris	So what was your favorite lab to do?
James	The candy lab. We got to do what we wanted to do with the candy, we got to mix our own ingredients, and mix stuff in. Our candy came out better than anybody else's.

The behavior of the students in the university labs demonstrated that incorporating their interests in the planning of the curriculum was a positive step toward improving the nature of engagement during laboratory activities. I wondered what distinguished these activities from those occurring in the classroom. The structure of the labs was one in which the students were working on concepts that they had chosen, all of the necessary resources were available, and the learning served a purpose that they could easily identify. The tapes of the labs showed learners who were using their creativity, intuition, and intelligence to pursue goals that were significant to them. Importantly, they seemed to be enjoying the doing of science.

From a sociocultural perspective, the structural characteristics of the learning environment (in this case, lab resources and equipment and topics chosen by students) and the practices of the students within the environment illustrate the dialectical relationship that exists between structure and agency. Sociocultural theory suggests that learning occurs within social systems and it is the structure of those systems that interconnects with the agency of the participants within those systems. A given structure can facilitate students meeting their goals or it can limit or preclude their attainment. In viewing the tapes of the student-suggested laboratory activities, I finally saw the energy, participation, curiosity, and natural intelligence that was not obvious in other day-to-day classroom activities. What had changed was the structure of the learning experience, and this structural change had afforded the agency of my students.

Although the labs appeared more successful, I had no method of assessing the nature of student engagement. The students looked as if they were doing "better" but how could one assess their actions? Interestingly, their practices in the lab did not seem to foster a greater understanding of the science content. When the students reported back to the class each Friday, the nature of their science talk was not improved. The structural changes of involving the students in the creation of the curriculum was a positive first step, however, it had not produced the quality of science discourse or construction of understanding that I had envisioned; moreover, I had no way of assessing the quality of their learning.

The opportunity to explore these issues presented itself, this time in a biology class, near the end of the school year. The latter weeks of May until the middle of June, is a time between final examinations and end of the year activities. It is a time that is traditionally open and involves less structured instructional activities. Once again, I asked the students for their suggestions of topics that interested them. However this time, I suggested that topics could be from any area of science. I introduced that particular change since, in my first class, the chemistry students had told me that restricting their choices to just chemistry had limited their selection of activities and they would have preferred to be able to choose any science topic. The students and I negotiated a series of activities that they felt would challenge and engage them for the remainder of the year. Students in all of the groups were responsible for a daily log and a final report that summarized what they had learned during the projects. The May projects included:

Chemistry of sugars and fats
DNA electrophoresis
Building rockets

Growing Fast Plants
Raising frogs from live eggs
Dissections: Comparative anatomy of different organisms
Building with K-nex: The roller coaster and Ferris wheel

One group of students undertook an investigation of the physics of amusement park rides by building a roller coaster and a Ferris wheel using a K-Nex kit. The projects explored the principles of conservation of energy and momentum and the relationship between potential and kinetic energy.

A second group of students was interested in ascertaining how forensic investigators used DNA evidence to solve crimes or parentage issues. They explored the process of separating DNA fragments using Agarose Gel Electrophoresis. They learned how to prepare and cast the gels into buffer solutions. Prepared DNA samples were centrifuged and delivered into the gels using micropipettes. Electrophoresis apparatus was used to separate the samples into fragments of differing lengths. Once the gels had run for the required length of time, they were removed from the buffering solution and stained using methylene blue dye.

Students interested in dissection compared the anatomy of different types of animals. Most students chose to compare frogs (amphibians) to grasshoppers and crayfish (arthropods) while some compared starfish (echinodermata) to earthworms (annelelida). Students picked organisms from at least two different taxonomic classifications and compared the respiratory and circulatory systems of those organisms to those of mammals.

The videotapes of these activities showed levels of student engagement similar to what had been observed in the labs done at the university; however, two groups, the "DNA crew" and the "dissection crew," displayed remarkable levels of motivation, perseverance, and creativity. In these groups, the nature of their conversations as they used the science content to discuss their observations and solve problems differed from the unimproved science talk described previously. The dissection crew consisted of students who were among the lower-achieving and usually less motivated students in the classroom, yet they remained focused and on task for many days with little encouragement or guidance on my part.

Jarvis was a member of the dissection crew. The nature of his science talk with others and me suggested that he had constructed a deeply personal and accurate grasp of the anatomical differences between mammalian and amphibian circulatory systems. To understand Jarvis's actions, I turned to Cultural Historical Activity Theory (Cole and Engeström, 1993; hereafter referred to as activity theory). The subject of this analysis is Jarvis; the community is the classroom, and the goal is the dissection of an amphibian heart. The rules are those negotiated by the teacher and the students; they detail the safe use of the dissecting tools, respect for the organisms being dissected, and traditional norms of classroom behavior. The group members determine the division of labor. The learning outcome is scientific fluency as evidenced by the use of tools, achieving the goal of dissecting a heart, and the quality of scientific discourse used to explain the activity.

Toward Scientific Fluency

It is an afternoon in late May. The students in the fifth period biology class noisily enter the room and move to their projects. We are well into the May projects and each group is in the midst of an activity. The work this day will be the completion of the Ferris wheel, the DNA crew's electrophoresis, and the dissection crew's continuation of their study of comparative anatomy. The DNA crew is running a DNA sample in a gel that they previously prepared. The dissection crew will divide in half. Half the group will dissect a grasshopper; the other half will dissect a frog. The goal is to remove and compare the hearts and lungs of these organisms. We establish where each pair of students will work and then set about distributing equipment and supplies. Kareem, Miaza, Nedwin, and Jarvis decide to dissect the frog, but later Miaza and Kareem move off to other organisms leaving Jarvis and Nedwin to complete the task. While I go to get gloves, Miaza and Kareem start to joke and flirt with each other. Jarvis however is not interested in their banter. He is concentrating on the frog and states, "I know all about this. Want me to show you all?"

Kareem and Miaza do not respond but Jarvis is not deterred; rather he stays focused on the frog pinned to the dissection pan. I arrive with three sets of gloves, which go to the two young women and Kareem. Jarvis is left without the necessary equipment. This is an early contradiction in the activity system, as Jarvis does not have access to necessary tools. Without the gloves his work is hampered. He says to me, "I need some gloves." There are no more in the classroom and so I have to leave to get gloves from an upstairs supply closet. Jarvis leaves the table and strays out of camera range, but as soon as I return to the room, he approaches me, gets the gloves, and moves back to the table. One of his partners has already started the dissection, but she is having some difficulty getting the heart out. Jarvis challenges his peers saying, "I bet you I can get the heart out." He takes center position in the dissection and begins the process of getting the heart out. Nedwin, Kareem, and Bruce (a visitor from another group) surround the dissection tray. Each one of them is helping; one holds the tray, another provides tools, and another is looking at the dissection guide. Bruce realizes that he is not wearing goggles and so he reaches for a pair. Jarvis taunts him "Got scared, got scared?" They all laugh at Bruce, and then continue to work. They are abuzz with observations about the amphibian, the color, the smells, and the size of the organs. One student disbelieves that there is any blood left in the animal, and another wonders if it's male or female. They continue to work and joke with each other, all the time focused on the amphibian's odd organs and the task at hand. At one point, Bruce bemoans the fate of the poor frog, the life that he had, and how he died. More laughter ensues, yet the focus of each person remains on the dissection tray; the heads remain bent, all hands are near or on the dissection pan, and there is no break in concentration. The community has spontaneously aggregated. I did not establish it, yet there is a very clear division of labor, the goal has been tacitly agreed upon, and they follow a clear set of safety rules which frame their collaboration and effective use of tools (sharp scalpels, probes, and scissors). Throughout the task they remain playful and energetic. When logistical problems arise, one student moves to the dissection guide or the computer, others adjust for his absence, and the dissection continues. Few obstacles arise. Finally, after about twenty-five minutes, Jarvis announces, "I got the heart out, Mr. Carambo." Bruce sings a song about the heart, and he and Kareem share a private joke and move away from the table. Jarvis and Nedwin do not move. At this point, Jarvis takes the heart from the dissection tray and states that he needs a magnifying glass and another tray. His goal has now become the dissection of the heart. He wants to compare it to the sheep's heart that was dissected several days earlier.

This goal is different from what we had previously discussed, since his personal goal has shifted and now requires an internal examination of the frog's heart. He needs another tray as his partner has continued on with an exploration of the frog's skeletal and muscular systems. A magnifying glass is needed because the organ is so tiny. Since I am involved with other students, Jarvis moves to the supply cabinet, retrieves the necessary equipment, and returns to the table. This is most impressive from a structural perspective because not only is Jarvis aware of the resources that he needs, but he is able to access them of his own accord and use them in the fulfillment of his own set of goals. At other times, the immediate lack of equipment might have created a contradiction or allowed him to become unfocused yet in this case he perseveres with the task. As he returns to the table, he passes in front of the camera and says, "Time to go to work." Jarvis works intently for another ten to twelve minutes, but then calls me over. "Mr.

Carambo I can't cut it 'cuz the scalpel won't cut it." He is having trouble since the dissecting tools are not fine enough, the heart is too small and he needs better tools. I provide him with a finer scalpel from an instructor's dissection kit and we plan out a new strategy. The students use the knowledge they have gained during the dissection of the sheep heart, along with their familiarity with the dissection tools to plan out their procedure and an associated division of labor. Jarvis will make the incisions and Bruce and I will serve as his assistants. The following is a transcription of our conversation:

Carambo	(*Looking at the heart*). [Re] member the sheep heart that we did. Can you tell which side's the left and which side's the right?
Jarvis	This (*gestures to indicate*) the bottom.
Carambo	That's the bottom.
Bruce	The top is (*gestures and indicates*)
Jarvis	This the left. (*gestures*) This the right. (*gestures*)
Carambo	Why is that the left side? Because it's, which side of the heart is bigger or smaller?
Bruce	This side is bigger. Because we pointed it out the other day
Carambo	Which side of the heart is bigger or smaller? If it's a pumping heart like ours?
Bruce	The right side's the biggest; the left side's the smallest.
Carambo	Well, which side pumps to the rest of the body?
Jarvis	What side? What side pumps to the rest? The left. The left side
Bruce	You talking about the ventricles right? The left ventricle is the one that pumps,
Jarvis	Yeah, it's the left
Carambo	So if it's the same as the sheep heart then maybe we could say this (*gestures*) might be the left side. Now how we cut? We cut the sheep heart like (*gesture*) like yo . . . we could cut it like this.
Jarvis	You wanna cut it down the middle
Carambo	No, I would cut it. Why don't you cut it? (*gestures*).
Jarvis	This how I'd cut it. (*gestures to indicate himself*) In my professional . . .
Carambo	You want to cut it down the middle?
Jarvis	We could cut it like down the side, like a book

Once the group decides on the procedure to use, Jarvis begins to dissect the heart. Bruce holds the tray and the heart in place, while I hold the magnifying glass. Jarvis continues to work at opening the heart. He identifies the top of the heart and orients it so that he can identify the left and right sides. As he opens the organ he sees what he considers to be blood. He remembers that earlier Bruce had doubted that blood could be present in the frog's heart. Jarvis announces, "I told y'all there was blood." The organ is quite small, but he and Bruce persevere. When he succeeds in opening the heart, he exclaims, "Got it! Got it! Yo, I'm the true . . ." I am called over to inspect. The heart has been expertly opened. The conversation immediately turns to whether it is chambered like the sheep's heart or different in structure. Jarvis tightens his gloves and exclaims, "That's what I'm about to find out." This is another self-generated goal. It is now nearly 2:30; they have been working continuously for over an hour with almost no teacher input or motivation. There have been few distractions; every time a goal has been reached, a new one has surfaced. During this last conversation, I become central to the table and the discussion. This happened accidentally, yet it is interesting to note how Jarvis reasserts his centrality. He takes the magnifying glass and probes out of my hands and regains control of "his" dissection. I realize his desire to assume ownership and I move to the side. Jarvis then continues to work alone. He does not need me. Once the heart is opened, Jarvis examines the chambers and identifies each one and relates their structure to the respective function of the frog's heart.

Jarvis	See (*pointing to the chambers of the heart with the probe*) there's two on your left and one on your right
Carambo	It makes sense to have two.
Jarvis	(*Still gesturing*) That's what I think.
Researcher	Can I see?
Jarvis	Come here, I'll show you. There's one right there, another one right there and there's another one right there.

It is now close to 3:10 and the period comes to a close. The crews begin to clear their workstations and the class ends. As we clear the table, Jarvis and I have a few minutes alone. We don't discuss the success of the day; rather we're looking at the frog's skin. It is at this point that I have a few minutes to "teach" about amphibians. Jarvis still has a few questions that he could not answer. He's curious about the difference between the frog and the sheep heart and why the lungs seemed so small. For these last few minutes I act like a teacher, and answer his questions and help him fill in his missing information. Finally it's time to go home so we stop and leave the rest for tomorrow.

Conversation with Jarvis

Jarvis's actions during the course of the dissection activities were those of a highly motivated and engaged student. His use of the dissection tools, requests for and use of additional resources, and his ability to compare and contrast the structure and function of the sheep's heart to that of an amphibian's heart were evidence of scientific fluency. While these events were encouraging to me as his teacher, they presented a disturbing contradiction because Jarvis's academic grades prior to the May projects were so low that he failed the year's biology class. He matriculated to his senior year only after completing a course in summer school. While this research suggests that building a curriculum around the students' interests fostered the use of their cultural and social capital to pursue relevant and important learning goals, Jarvis was unable or unwilling to invest his social and cultural capital to the same degree during the year's regular course of study. In order to understand why this occurred, I invited Jarvis to view the videotapes of his activities in the dissection crew and share his perspectives on himself as student, on school, and on the role of teachers. Following are excerpts from our conversation.

Carambo	On this day, you came into class at 1:30 and you didn't move 'til 3:09. Do you see yourself as being different on this particular day?
Jarvis	No, I see myself as the same person; I don't see myself different.

Initially, I found it odd that Jarvis would see no difference in his practices throughout the semester and those enacted as a member of the dissection crew. In retrospect, I understand that Jarvis was viewing himself, as he knew himself to be, taking account of all of his social and cultural capital. While dissecting, he was funny, cooperative, enjoyable, sociable, and very playful—qualities that often served to take him away from learning goals as he pursued other more important personal goals. During the dissection activities however, all of these (historically distracting) qualities were at the service of a learning goal that had captured all of his faculties. To the outside observer, it looked as if a different kind of student was involved. The reality was that Jarvis had found an environment in which he was free to be agentic. The Jarvis that I had seen in class throughout the year appeared to be a disinterested, disengaged, and failing student. I reminded him of his practices in the regular class, as I still needed to understand why the dissection activities had such a significant impact on his behavior.

Carambo	I remember that in this class, sometimes you would mess around a lot and you would get drawn off, and you'd be hanging out with people and goofing around, but this day I didn't have to say anything to you, and every time you had a chance to goof off you didn't.

Jarvis	This [the dissection] was different because I was real, real interested in this right here.
Carambo	This really interested you? Did you always want to do dissections? Did you always want to cut things open? What was so interesting about this?
Jarvis	For real? You, like how you was askin' everybody what we wanted to do and everybody was suggesting and I was like the only one out of, like I was the only one out of four. It was like me and three other people out of all the classes that wanted to do dissections and you agreed and we were talking about it, and I was very interested and like I couldn't wait. This right here? This really interested me.

It is still not clear exactly what component of the activities captured Jarvis so deeply but he suggests in the following exchange that it was his ability to pursue a topic of his own choosing, and the respect that was accorded to his interests that are important to him and his peers.

Carambo	If you could give me some advice on how to make classes good for students, what would you say? What would you do? How would you make school more interesting? What would you do?
Jarvis	It's got to do with the teachers. . . . It's all right for the teacher to come in and get straight down to business, do whatever they got to do, do whatever he's supposed to do to teach the kids, but sometimes you got to get the kids' opinions, 'cuz that's how we go to sleep in the class and do all unreasonable stuff.
Carambo	'Cuz we don't listen to you?
Jarvis	Yeah, but it's not that, but (*pause*) I'd say if we was you know, like we got to work with each other, we could come up, you know, if we meet halfway. We can come up with an agreement, and that way everyone would pass the class.

This advice was not unexpected, however the reality is that teachers are not always able to incorporate students' interests into their lessons. I asked Jarvis to address this issue.

Carambo	So find somewhere in the middle, [meet halfway] what happens if we're in the middle and we can't do what you want to do? What would happen then? How could we make it work ok then?
Jarvis	If you can't do [all] what we want to do then you can put a little, you know you could put a little bit in your lessons that interests us, about our suggestions that we gave to you, that way teacher and student could communicate and they have a better relationship, you know between the student and teacher, then the student would want to do his work.

Carambo	Would that make you feel more like a partner in the whole thing? If we could put a little bit into our lessons? How would that make you feel?
Jarvis	It would make me feel more interested in the work.

Jarvis is reiterating an earlier point. While the inclusion of student interest is important, listening to students fosters an atmosphere of mutual respect that empowers them and lets them know that they are important members of the learning community. However, power inequities within most urban schools are such that students have little power to be involved as curriculum developers.

Carambo	Do you feel like you have no power or control over what goes on in class? How do you feel?
Jarvis	You really don't have no power, you know, some people, some teachers say the teachers don't even have no power, they got to go by what they get, that's like true, but they still don't go, some teachers don't go by the lessons that they have to give to us, they do whatever they want to do so, why can't we both communicate with each other and maybe you know class would be a lot better?

Pondering the Contradictions

This conversation took place during Jarvis's senior year. He had attended summer school and was optimistic of his chances of graduation. He told me of plans to attend college and pursue a well-paying career. Although his actions in our May projects showed a promising, articulate, and intelligent student, during the next year he persisted in the kinds of actions that had caused him to be a failing student in my biology class. His senior year was marked by suspensions, truancy, and many failing grades. He was therefore unable to graduate with his class and must attend evening classes in order to graduate from high school.

One might conclude that the events that befell Jarvis were due solely to his deficiencies as student. My experiences with him suggest otherwise. One can only wonder what might have occurred if students like Jarvis were to learn in classrooms that allowed them to utilize all of their rich stores of capital in challenging the status quo and establishing productive learning environments.

Beyond Simple Solutions

By engaging sociocultural lenses, I learned that the contradictions of student disengagement in my science classroom could be resolved by including the interests
of my students in the curriculum. The use of activity theory provided a framework to review students' practices and study the classroom as a place where multiple goals could be met. From this perspective I allowed students the space to
employ their rich stores of social, cultural, and symbolic capital as resources for
learning science (Seiler, 2001) and provided students greater autonomy to identify and use appropriate tools, divide labor equitably, select tasks that aligned
with their interests, access human resources they regard as appropriate, and use
their own language along with canonical science concepts to communicate and
represent their learning.

The quality of the participation in the May projects was higher than I had
observed in the lab activities undertaken at the university. This may have been
due to the greater range of possible choices provided in the May projects since
students were not limited to choosing projects from just one scientific discipline.
It should be noted that the success of these projects is not attributable to students
engaging in interesting "hands-on" activities, since we routinely did such activities. Instead, the success may be due to the genuine respect that was afforded to
their ideas and the boost this gave their self-esteems. Jarvis's comments tell us
that students are empowered when they become active partners in the creation of
their curriculum. This empowerment increases their social capital, positively impacts their sense of self, and affords them opportunities to incorporate their vast
amounts of social, cultural, and symbolic capital into learning. Jarvis draws attention to the significance of the quality of the relationship, the mutual respect
that existed between him and me, respect that allowed him the freedom to incorporate parts of his person that he traditionally kept hidden. This suggests that it
is the quality of the teacher-student relationship, and not merely the doing of
hands-on activities that affords greater engagement and infusion of student capital into the classroom. It is as Jarvis says:

> If you can't do what we want to do then you can put a little, you know you
> could put a little bit in your lessons that interests us, about our suggestions that
> we gave to you, if you can't take our suggestions and do whatever we say, you
> could take some of our suggestions and put it in your lesson, that way teacher
> and student could communicate and they have a better relationship, you know
> between the student and teacher, then the student would want to do his work.

It is my hope that by listening and incorporating the voices of our students into the enacted curriculum, we can create the kinds of relationships that allow students like Jarvis the opportunity to be successful learners.

Editors' Perspectives

Jarvis typifies so many males we have encountered in our studies of urban science education. He showed his competence in science, persistence, and intense focus and interest in a dissection activity that he had suggested to his teacher. In the context of that activity he enacted science fluently, in terms of psychomotor skills involved in dissection and canonical representations of the structures and functions of the hearts of different organisms. For as long as he was involved in dissections Jarvis was the "true," providing evidence that his identity was impacted by his success at doing science. He argued, persuaded, encouraged others to work with him, and finally, surpassed the group's goals. In one activity Jarvis was a success and his accomplishments were as impressive as high school counterparts in suburban and urban schools. Yet those high quality attributes were not consistently shown and Jarvis rarely showed the same degree of focus and fluency in science. Indeed like so many urban youth with whom we have collaborated Jarvis failed his science course and struggled to succeed in high school.

As Roth showed in chapter 4, it is unlikely that we resolve contradictions like those raised above if we only study Jarvis's social life in the fields of science education and the school. Carambo's selection of a dissection activity was a source of symbolic capital for Jarvis. Hence, science educators are advised to listen to students and incorporate their suggestions as signs of respect for their interests and ideas. However, why does Jarvis place low value on school achievement and how can achievement be measured authentically to assess what Jarvis can do and knows of science? Also, what social and cultural lenses are needed to identify the capital that Jarvis possesses and enact a curriculum so that he can fully apply his capital in ways that are analogous to his participation in the dissection activity?

As we have exhorted elsewhere, cogenerative dialogues are likely to yield rich cogenerated outcomes that might alter the nature of science education. As LaVan and Beers have shown in their research (chapter 8), cogenerative dialogues are ideal for discussing participation in a number of fields that intersect or are nested within the school field. It is likely that participation in the neighborhood and home produces goals and values that are at odds with science education being perceived as likely to improve the quality of social life for students like Jarvis. Accordingly, we regard it as a priority for science educators to learn more about the lifeworlds of urban youth and the social, cultural, and historical constituents that shape their identities and the extent to which they value science education.

Chapter 10

Female Sexuality as Agency and Oppression in Urban Science Classrooms

Melissa Sterba

Sexuality is one of the oldest forms of power for women. It has been used since the beginning of time and in so many ways. It's no wonder that these girls use sexuality. It is and will probably always be about power.

—Cristobal Carambo, City High

Female Sexuality as Power

Urban science classrooms reflect many of the contradictions women face in society at large. Many educators and administrators find that students continue to focus on appearance and its effects on peer interactions more than on learning (Alexander and Alexander, 2001). Schools have responded to this with policy interventions, such as the implementation of dress codes intended to control overt displays of female sexuality that could potentially disrupt classroom dynamics. A teacher at City High elaborates, "I wish that my girls could just concentrate on learning as much as on how they look and who they get to pay attention to them." Instead of focusing on learning science, some female students focus on cultivating sexuality as a resource that can be used to gain power and control in their lives. A tenth grade student from a chemistry class explains, "If you nice to look at, you go places with life. What we do in school don' matter in no way." Many of the interactions occurring in urban science classrooms indicate that physical attractiveness and female sexuality can be accessed and appropriated as a resource to reach personal goals and in particular, to gain the respect of peers. Yet, at the same time, closer analysis of these interactions reveals that relying on female sexuality can simultaneously prevent the acquisition of respect and prevent women from learning science.

This chapter examines the structural inequalities related to female sexuality that some female students face in urban science classrooms and in society at-large. For this reason, I focus on the experiences of Rashida, a tenth grade student who has acquired a "street name" by purportedly having indiscriminate sex with multiple partners. In many ways, Rashida fits the overall description of City High. She is a seventeen year-old African American female student who, despite much skill, has not met her academic potential. She is frequently absent from school, and has already failed tenth grade once. In class, she often sleeps and does not pay attention. Instead, she focuses on gaining the respect of peers, often by using physical attractiveness and in particular, sexuality. In this chapter I utilize multiple data sources from math and science classes collected over the academic year 2001-2002, including interviews, videotapes, and field notes to analyze Rashida's interactions at school. At times, I use microanalysis to reveal nuances in peer interactions involving Rashida. Given the data, I argue that many female students have experiences similar to Rashida, especially at urban high schools such as City High.

Maintaining Respect

Sexuality remains closely tied to respect for women. This is especially true when addressing the stereotyping of minority women, and especially African American women, as highly promiscuous and sexually aggressive (Collins, 2000). Previous research and policy interventions perpetuate such misconceptions about the sexuality of minority women living in inner-city areas such as West Philadelphia. These often fail to take into account the sociocultural context as well as the continuing forces of race, compulsory heterosexuality, class, and gender oppression in the lives of women today (hooks, 1995). Thus, when examining female sexuality, sources of both agency and oppression must be considered. Although agency can be conceptualized in many different ways, for the purposes of this chapter, agency is defined as "the extent to which (women) are able to act in ways that allow them to be in charge of their own lives or to exert control over their social position" (Seiler, 2002, 14). Similarly, oppression refers to the extent that women are prevented from improving their lives, often by structural inequalities, and at times, institutional practices.

Today, conceptions of respect, as well as the struggle to earn it, reflect societal structures of inequality as well as shifts in the symbolic and cultural elements of everyday life (James and Sharpley-Whiting, 2000). This is particularly true since we live in an era of gender transformation (Walby, 2000) where the structural inequalities related to gender are dynamic, mutable, and fortunately, open to change. Thus, as young women such as Rashida struggle to gain respect in urban classrooms, they simultaneously provide opportunities for transforming or reinforcing structural inequalities. Their actions can be both a source of agency or oppression. For these reasons, it is vital for science educators to develop a greater understanding of many of the interactions occurring in their classrooms, and how these interactions are intimately linked to the acquisition and maintenance of respect by young women.

The Female Code of the Street

For both female and male urban youth, a "code of the street" exists that shapes how respect and power are earned in an environment where lack of economic and educational opportunities, isolation, racism, and institutional indifference can make individuals feel powerless and disrespected (Anderson, 1999). At City High many students focused on socializing and maintaining the respect of peers. Often, they used schemas and behaviors associated with a "street" youth subculture rather than those associated with school to acquire respect. This was especially true in relation to female sexuality, where many female students dressed to highlight physical attractiveness and engaged male students in sexual banter in the classrooms at City High. At times, female students used physical aggression to defend their sexuality. Of course, these actions, while seen as untraditional or at odds with white, middle-class conceptions of femininity for women, are not new. Previous research documents their prevalence. For example, Anderson (1999) describes:

> Increasingly, teenage girls, most often associated with the street, become involved in group and individual fights. In many ways their fights are not unlike those of boys. Their goal is often the same—to gain respect, to be recognized as capable of setting or maintaining a certain standard. They frequently try to achieve this end in ways that have been associated with young men, including posturing, abusive language, and the ready use of violence to settle disputes,

but the issues for girls are usually different. Although conflicts over turf and status exist among girls, the majority of the disputes seem rooted in assessments of beauty (which girl in the group is the "cutest"), competition over boyfriends, and attempts to regulate other people's knowledge and opinions of a girl's behavior or that of someone close to her, including friends, siblings and parents. (63-64)

Yet, Anderson's description does not go far enough. He fails to discuss how emerging street behaviors indicate the structural evolution of gender and more specifically, femininity in relation to race, class, and compulsory heterosexuality. In other words, while the women's rights movement secured many victories, they are often full of contradictions, particularly when coupled with an analysis of structural forms of power and oppression evident in urban science classrooms today.

The Dialectical Relationship between Men and Women

Everyday, in countless urban science classrooms, men and women struggle to earn respect alongside each other, in a dialectical relationship. How men struggle for respect affects how women struggle and vice versa. Thus, social constructions of femininity and masculinity are not binary categories of social existence. Notably, the traditional conceptualization of femininity and masculinity as binary opposites contributes to the categorization of men and women in terms of their difference from one another (Collins, 2000). Such binary, dichotomous thinking shapes the understanding of human difference in oppositional terms, obscuring the overlap of the symbolic and cultural elements shared by women and men. In everyday life, however, women and men utilize many of the same resources to define femininity and masculinity.

Despite the persistence of gender inequality, many of the cultural constructions of respect are similar for adolescent African American women and men. Specifically, in urban areas such as West Philadelphia, both females and males gain respect through rapping, athletic prowess on the basketball court, the acquisition of fast money, and expensive clothes. Applying Sewell's theory of agency (1992) to Anderson's *Code of the Street* (1999) shows that through the exercise of agency, urban women may borrow some of the symbolic elements of respect from men, and transform their meaning through appropriation. Thus, while urban men may use sex as a method for gaining the respect of other men, so too may urban women use sex as a method for gaining the respect of other women and at times men. To that effect, some female students in urban science classrooms dress provocatively and approach male students with sexual innuendo and gestures. For example, Rashida often wears form-fitting, black stretch jeans to school and moves her hips to accentuate her well-formed figure as she passes young men on her way to her seat at the start of class. Further, she often captures the attention of young men, who look up from their desks and comment on how "good" she looks.

The examination of the use of female sexuality and in particular, female sexual performativity shows that femininity and masculinity are dynamic, evolving cultural constructs. Likewise, it provides evidence of the perpetual change of the symbolic elements of everyday life, often through the agency of individuals. This is because cultural schema are largely *generalizable and transposable* (Sewell, 1992), allowing women to transform current constructions of femininity by borrowing elements of masculinity. Thus, depending on context, sexual performativity and aggression are resources that can be used by women to gain respect. Adolescent females, then, can use the sexuality and aggression that once defined traditional conceptions of masculinity and, in so doing, they can create their own "code of the street" (Anderson, 1999). Given the dialectical relationship between structure and agency, they can borrow the cultural schema of urban men, and as a result, challenge and at times, possibly transform traditional constructions of femininity and masculinity. Further, their appropriation of the resources and schema once associated with constructions of masculinity may even question deeply entrenched inequalities relating to race, compulsory heterosexuality, and class.

Challenging Deficit Views of African American Female Sexuality

Currently, many adolescent African American females struggle to re-define femininity. In large part, this re-definition of femininity requires adolescent African American females to dispel stereotypes and persistent images of African American women as mammies, matriarchs, welfare recipients, and hot mammas (Collins, 2000). Clearly, challenging these images has long been a central component of black feminist thought. Many young women in urban science classrooms also challenge these images as they earn respect and, perhaps unconsciously, attempt to transform structural forms of inequality. However, structural inequalities remain, especially in relation to the image of "hot mamma." Many young women, such as Rashida, do have street names and at times, perhaps unconsciously, they draw on deficit images of African American women as they struggle to engage in self-definition. Rashida explains, "There ain't nothing wrong with usin your body if it nice." However, in the end, relying on female sexuality and defining oneself as a "hot mamma" can lead to a lack of respect from both female and male peers.

One Hot Mamma: Rashida's Attempts to Earn Respect

Rashida, like some of the women at City High, is a "hot mamma." She relies on her sexuality to earn respect. Unfortunately, Rashida fails to accumulate the respect of peers. Instead, she has a "street name." A young woman acquires a "street name" by making herself available for sex to young men, often indiscriminately. Ivory, a peer of Rashida's at City High, describes the behavior this way, "They be havin' sex with everyone." Rashida has a reputation among her peers of being promiscuous and incapable of saying "no." Yet, when told from Rashida's perspective, all of the females at City High have sex. Moreover, they have sex with multiple partners. "It's like everybody be doin' it all over this place." Still, while many of her peers may be engaging in sexual activity, there is a reason why Rashida has a "street name" and they don't. Namely, Rashida does not appear to be in control of her sexuality. Thus, men perceive her as a sexual object to be taken, while women perceive her as a threat to meaningful, monogamous relationships.

Over the past two years, I witnessed both male and *female* students harassing Rashida in school. At times, they called her names, such as whore, bitch, and pussy before letting her pass. On one occasion, Rashida was late for class, despite trying to arrive on time. As she attempted to enter the room, Shakeem stopped her at the classroom door, blocking her entry. Looking her in the eye, he stated, "Move, bitch! Get out my way. Get out my way. Get out my way. Move, bitch!" While the dialogue on the videotape of this interaction cannot be heard, I recorded it in my field notes at the time. However the video does capture Rashida's body movements and struggle to escape Shakeem's grasp. Specifically, the video shows Rashida approaching the doorway to the classroom carrying her books. As she approaches, Shakeem moves away from two male friends talking on the left-hand side of the doorway. Rashida is smiling at Shakeem. Shakeem grabs her right wrist, clasping it firmly. Instantaneously, within two frames of video footage, Rashida is no longer smiling. Rather, she stops in the middle of the doorway. Shakeem says something to her. As I approach the door, I hear, "Move, bitch! Get out my way. Get out my way. Get out my way. Move, bitch!" Rashida pulls out of his grasp and hurries to her desk at the other side of the room, out of the camera's range. Shakeem turns back to his friends, smiling.

The entire incident occurred within six seconds, broken up abruptly by my approach.

Although the duration of this incident was short, Rashida felt its effects throughout the lesson. She sat at her desk with her head resting against her arms, seemingly asleep. Although sleeping through class is not unusual for Rashida, after the incident she went to her desk and immediately cradled her head against her arms, closing her eyes. Unfortunately, there are many times when Rashida, like many of her female peers, suffers such humiliation before entering a classroom to learn science.

During my two years at City High, I became accustomed to male students pursuing females at school. As Anderson describes the male "code of the street," males try to work their "rap" in order to have sex, usually by talking smooth, complimenting a female, or stressing what they can offer in terms of appearance, clothes, money, and status (Anderson, 1999). At other times, males touched or fondled females. On very few occasions, they even threatened females physically. The use of sexual aggression by males centered on whether or not they perceived an individual female as an object of sexual conquest. However, once identified by men as objects, these women also lost the respect of female peers. Thus, once targeted by males as having a "street name," a female student such as Rashida suffered ridicule from both male and *female* students. Not only did males proposition, touch, and harass women like Rashida, but females did too. Ivory explains.

> It's not just girls havin' sex. It's that they have sex and got a street name so they, they can't control they reputation no more. They be doin' it with everyone. Once you got the street name, then there never no more respect for you.

The lack of respect Rashida receives from both male and *female* peers is evident in everyday classroom interactions. Videotapes revealed several instances of Rashida being called names, fondled, and at times, physically threatened. Further, several examples of females using physical aggression against Rashida surfaced. For example, one clip begins with Dawn, a tenth grade female student, and Rashida sitting side by side in Chemistry class. Dawn is on the left-hand side of Rashida. Dawn is banging her fists softly on her desks in rhythm, rapping quietly as the lesson continues. After four seconds, her right fist stops midair and reaches over to Rashida's desk. At this point, she bangs her fist in rhythm against Rashida's desk. She repeats this action twice. Rashida is smiling. Dawn then grabs Rashida's left wrist with her right hand, and Rashida, still smiling, moves her desk slightly forward, escaping Dawn's grasp. Almost instantaneously, within five frames of film, Dawn inches her desk forward and continues to rap her fist against Rashida's desk. Rashida sits there for approximately three seconds until Dawn grabs Rashida's pen out of her hand. At this point, Rashida immediately pulls her desk several inches forward while attempting to pry the pen from Dawn's hand. The student teacher walks toward the girls, returns Rashida's pen, and ends the interaction.

At first glance, this interaction appears to be a minor use of female physical aggression. Yet, it is one example of many such one-sided interactions between Rashida and Dawn. Moreover, this incident shows that although such interactions appear to be in jest, they are disruptive of the learning process. Rashida, who has already failed tenth grade once, focuses on avoiding Dawn rather than on learning chemistry. Further, this interaction demonstrates that there is a point where Rashida begins to take the aggression seriously. She stops smiling, and rearranges her seat to avoid Dawn. The actions accelerate and ultimately, a teacher intervenes.

In order to better understand the use of physical aggression against Rashida, I showed the video segment to four of her classmates (two females and two males). Not only did they immediately recognize the clips, they laughed about them.

Melissa	So, why is this so funny? I mean, (Rashida) isn't really doing anything, right? So why is she picked on? . . .
Shakeem	She got a name for herself. She be pumpin' a lot of johns, you see what I'm sayin'?
Melissa	Is that why (Dawn) does things like that to her? Is that why she picks on her?
Ivory	Yeah, that part of it. She ain't got no respect. She lie to her and so Dawn sometime do a head dunk.
Shakeem	(laughing) BAM!
Melissa	Doesn't that hurt?
May	It do, but she deserve it.
Ivory	Can't take care of herself no way. It a shame.

I asked Rashida about how peers treat her.

Melissa	Why do you let them do that to you? Why do you let them touch you and call you names like that?
Rashida	I don' know. They jus' do it like that, that's all.
Melissa	But I mean, don't you want to do something about it, when they call you names?
Rashida	It don' mean nothin to me. It jus' the way it is here.

Thus, by earning a "street name," Rashida expects the use of physical aggression against her, from both males and *female* peers.

The use of female physical and sexual aggression is not confined to the classroom. In particular, female students can use physical and sexual aggression in areas of the school where teachers or authority figures are not present. Such places make it easier for female students to engage in the "code of the street." At City High, the commons on the second floor is an area where students interact and pass each other between classes. Often, they are out of the view of teachers. On a videotape showing the passage of students through this area between classes, a young man approaches a young woman. He whispers something to her, and attempts to touch her in a sexually suggestive manner. The young woman grabs his hand and pushes him away. He stumbles back to his friends. Forty seconds later, the same young woman comes back at the man, now talking to his friends. She walks behind him, reaches around, and hits him across his chest. Next, they struggle as she tries to trip him in front of his friends. Although dialogue cannot be heard, it is apparent that the male's friends are laughing at the interaction. Unlike Rashida, this female student succeeded in defending her sexuality through the use of physical aggression against a male student. She was able to use the "code of the street" to defend her sexuality and earn the respect of peers. Through her actions and by taking control of a situation where a male made seemingly unwanted sexual advances, she was able to show that she is more than just a "hot mamma."

Finding Female Agency in a Man's World

To the young man the woman becomes, in the most profound sense, a sexual object. Her body and mind are the object of a sexual game, to be won for his personal aggrandizement. Status goes to the winner, and sex is prized not as a testament of love but of control over another human being. The goal of the conquest is to make a fool of the young woman. (Anderson, 1999, 150)

Despite the agency of many female students in urban classrooms, sexism exists. That is, men often try to gain respect by proving their ability to enter into sexual relationships with female peers. At the same time, women must defend their desirability from unwarranted sexual advances in order to acquire and maintain respect. As the incident with the young woman in the commons illustrates, female students must exercise agency and appropriate resources, such as the physical and sexual aggression once associated with masculinity, to counter unprovoked advances and control the situation. Still, when women use the resources typically associated with masculinity, their actions are sometimes portrayed through a deficit lens. Their enactment of "street culture" stands in stark contrast with the dictates of school culture. As a result, many educators and policymakers continue to worry about the behavior of young women such as Rashida, especially when these behaviors cause disruptions in the classrooms.

Sexuality, which for men emphasizes having sex and for women the denial of sex, continues to shape how women struggle to gain respect and power in the classroom and in society at large. This is because the ideology of femininity is powerful, seen as a natural part of everyday life for many women, and taken for granted (Moi, 1999). However, ideology is slowly challenged. As a result, the ability of males to take control of female sexuality and heterosexual relationships weakens gradually over time and in certain contexts. Today, females often struggle alongside men to take control of sexual identity and activity as a method for accruing respect. However, the rules of the struggle for respect are not as clearly defined for women, and the consequences remain much more immediate, as in the case of pregnancy.

Sex can earn respect on the streets of Philadelphia. For males, engaging in sex without establishing a long-term relationship based on commitment is a sign of "street smarts," a pillar of male respect (Anderson, 1999). Even today, many men engage in sex without having to live with the consequences of early and/or unwanted pregnancy. In contrast, many females at City High recognize the sacrifice and hard work of having a family. They question female subordination to men. Yet, traditional middle-class, white norms of femininity, motherhood, and family remain, often shaping the motivations, aspirations, and everyday lives of some African American female adolescents in urban areas. For these females, respect often means finding a man to provide for them, financially and emotionally. As a result, many women construct their identities in terms of their ability to give men children as a sign of love, affection, and commitment. Often, they hope men will offer marriage, financial assistance, and a sense of personal accomplishment. The young women also see the child as a source of unconditional love. Thus, as males on the street seek to engage in sex to show their manhood without having to be tied down by commitment, some young women seek to prove their womanhood by having children in order to solidify stable, meaningful relationships or simply to "feel good" as Rashida put it. While men seek avoidance of stable relationships, some young women seek the very same stability men wish to avoid (Anderson, 1999). Inevitably, cultural struggle ensues and young women, like Rashida, must negotiate powerful cultural contradictions.

Rashida knows about taking care of babies, even at the age of seventeen. She complains of having to watch children for friends and family, especially for the hair stylist who works in the basement of a friend's house. Still, Rashida expects women to be left with childcare duty. For example, she was not surprised when the two male undergraduate students from the university working in her geometry class did not know that Similac was used to feed infants in place of breast milk. Indeed, Rashida continues to see females as having primary responsibility for children, emotionally and economically. Having children, without the support and active involvement of a male figure, is part of Rashida's everyday life and experiences.

Despite the challenges associated with having children while young, Rashida continues to utilize her sexuality for attention and engages in sexual activity. For Rashida, sexuality is a resource that she can access to accumulate respect. This is true despite a scare Rashida had with pregnancy. In February, Rashida asked to speak with me privately. The following is an excerpt from my field notes.

> I spoke with Rashida in the hallway before lab today. Rashida thinks she is pregnant. She is a month late. While she hasn't always been regular, she is scared. I am unsure of how to advise her. Obviously, she needs to find out whether she really is pregnant before she jumps to conclusions. I hope this is just another "scare." Given her prior work record, I do not think she will graduate from high school if she has a baby to take care of while finishing her education.

While it may seem a contradiction that Rashida can simultaneously resent having childcare responsibilities and engage in unprotected sexual activity, many women find themselves in similar situations. Some of Rashida's peers describe young women having unprotected sex and ending up pregnant, only to have to take of the child without the support of the father. Specifically, May, a classmate of Rashida, relates this example.

> And then like one of my friends, I know that she had a kid, because she was pregnant last year when she left. Um, she wasn't sure, but she wasn't like upset or anything. She was like, um, I think I'm pregnant. And her boyfriend wasn't sure, but he's hopin' she pregnant. I was like, you pregnant? What you gonna do? She like, I don' know. She like she work out for next year. So then she back with this big stomach and I was like I know you all pregnant. And then she had a girl. And now he always make her um, he make her carry the baby. He don' wanna hold that baby. He won't even hold the stroller while he jus' bein there. He make her carry bags. No. He don't do nothin' for that baby. And that his baby too.

| Melissa | I guess that takes an awful lot of effort to look good. Why do you do it? I mean, why spend all that time on your hair and makeup if you're just going to change it anyway? |
| Rashida | I look good. I like to look good. That way people notice me. |

Rashida dresses for attention. She frequently changes her hairstyle, emulating 'Lil Kim, a popular rap artist. Like 'Lil Kim, Rashida shows off her feminine assets and sexuality by wearing tight-fitting and sometimes revealing clothes. Like some women engaged with the "code of the street," Rashida relies on her sexuality as a source of attention and the possibility of respect, even if at times it leads to discomfort in the form of being propositioned, fondled, or targeted for physical aggression. In the end, Rashida's reliance on female sexuality, as embodied through style, provides potential for both respect and disrespect simultaneously.

For some young women, like Rashida, clothing and style provide powerful evidence of the shifting cultural and symbolic constructions of femininity. Rashida describes, "There ain't nothin' wrong with showin' what you got if it look good to people." For Rashida, her dress and body capture attention and appreciation, from both males and *females*. Notably, Rashida recognizes that if males appreciate and value her body, females will too. Likewise, she recognizes the powerful currency of the "code of the street" on the streets of Philadelphia and in society in general. She sees that many women from similar circumstances, such as rap artists 'Lil Kim and Eve, use revealing clothing in order to gain status, wealth, and respect. Many of her peers confirm Rashida's observations. While watching rap videos on Launch.com, May explains.

| Melissa | So why are they all dressed like that? Why do you think they do that? |
| May | They got to. To get some respect in that business. It like men control rap so women do what they got to, to make it. |

For young women who both listen to and produce rap, female rap artists' styles provide powerful examples of how women can earn respect. Thus, if Eve, who is from Philadelphia, can achieve fame and fortune through revealing dress, Rashida feels she can too. After all, she does not believe that school will make a difference in her life: "I don' really learn here anything that matter. It boring." For Rashida, the effects of sexism may have left the realm of conscious awareness to become part of who she is. She accepts the "code of the street" without conscious reflection of the contradictions it presents in her everyday life. Unconsciously, sexism permeates Rashida's life, in her movements, dress, tastes, and personal aspirations, and without the support of strong females who are able to navigate many of the contradictions within the female "code of the street," she is left with few alternatives.

Struggles for Respect

In countless urban science classrooms, female students struggle to gain respect. Often, their struggle has little connection with teaching and learning and instead relies heavily on the "code of the street." To that effect, female students sometimes access and appropriate sexuality, on both conscious and unconscious levels, to gain attention, companionship, and ultimately, respect. Still, female sexuality remains a double-edged sword. A contradiction emerges where females who rely on their sexuality, such as Rashida, are seen as "hot mammas," or as sexual objects to be taken. They are subject to ridicule, humiliation, and even physical aggression by both male and *female* peers. In the process, they lose the respect of peers and the opportunity to learn. Ultimately, the experiences of female students like Rashida show that resistance to schema and practices associated with sexuality is not an easy process for young, inner-city, African American women. It is a process scarred by contradiction and pain.

Editors' Perspectives

Rashida certainly had the intellectual capacity to succeed at chemistry and when members of our research team made an effort to connect chemistry to her life as a female she sparked up and showed more interest in learning. So, when we made an effort to interact with her about science topics in which we felt she would be interested she responded in ways that produced successful interactions. However, these moments of interest were not transferred more generally to having an interest in science or participating in a curriculum that she perceived to be of little or no value. This is a dilemma for science teachers because it was apparent that her communication skills were advanced and she could easily have been one of the highest achieving students in her class. Like Jarvis in the previous chapter, Rashida showed that she could be fluent in her enactment of science, but did not regard persistent participation as a goal on most occasions that science was scheduled. We regard it a priority for science educators to find ways to include in the curriculum a steady diet of topics that are central to the interests of urban youth and relevant to their lifeworlds. As Carambo showed with Jarvis, and Seiler and Beers with other students at City High, paying attention to the students' interests is a sign of respect that their interests matter and can catalyze periods of intense engagement in activities associated with their interests. Is it possible for science educators to infuse into each class period some time for students to participate in science that relates to their interests? The payoffs might be considerable for students like Rashida.

The physical interplay between males and females that Sterba describes in class, the hallways, and the commons might be an example of *playin* (Elmesky, chapter 5) and students might use the presence of many students in close proximity as a structure that affords their attainment of other goals through exchanges of capital as they interact with one another. For example, instances of males groping females such as Rashida occurred often enough to constitute a pattern that was condoned by her and set up a structure for others to show their disrespect for someone who had earned a street name by engaging in similar practices. Groping thereby contributed to a cycle in which students could show their disrespect of Rashida and thereby earn the respect of their peers. Such a pattern had deleterious consequences for Rashida whose identity was shaped by frequent sexually explicit actions and comments laced with sexual innuendos from males and sarcastic and sometimes physically violent interactions with females from her peer group. Interactions of this type were disruptive to Rashida's learning and a source of constant negative emotional energy. Similarly, the opportunities for sexual byplay may have been a constant distraction for some of the males in the class, especially those who sat close to Rashida. Science educators need to address issues of sexuality and oppression in urban schools and create structures that allow students to focus on learning without being constantly distracted by sexually explicit interactions.

Chapter 11

Meeting the Needs and Adapting to the Capital of a Queen Mother and an Ol' Head: Gender Equity in Urban High School Science

Kathryn Scantlebury

The principal of City High, herself an African American female, recognizes the interconnections between gender, race, and socioeconomic status when she notes that being black, poor, and female means that the girls in her school have three strikes against them. Thus there is a great need for contextualized research in this area, especially as it pertains to African American female students in urban schools. African American women are subject to the embodiment of racism and sexism, which is often used to ignore the class/economic issues that also impact their lives (Collins, 1998). Sterba (chapter 10) identified multiple ideologies that impact the identities of African American girls in potentially hegemonic ways. Although African American parents support gender equality for their daughters in the public sphere, they often have strong stereotypical attitudes toward their daughters' roles in the private sphere. Hence, expected gender roles for African American girls are a combination of traditional feminine roles with new images of economic independence, assertiveness, and community activism. This chapter foregrounds two females from Cristobal Carambo's chemistry class (chapter 9), exploring their practices in this science class and in their lives outside of school.

In this chapter I examine the schema and practices they enact and the social, cultural, symbolic, and economic capital they represent. In doing so, I identify and describe resources and contradictions that are not immediately obvious.

In Carambo's chemistry class, there were more boys than girls, which is unusual because the attendance of males in urban schools is often truncated by high rates of truancy, drop-outs, and incarceration. However, the different foci of the school's small learning communities (SLCs) led to a school-wide uneven distribution of girls and boys that reflect gender stereotypes. For example, the Health SLC was predominantly female, while Science, Education, and Technology (SET), the SLC in which Carambo taught, was predominantly male. However the higher number of boys enrolled in SET reflects boys' interests in sports not science, since the previous year this SLC was called Sports, Entrepreneurship, and Technology and males comprised about 75 percent of the students in SET. The remainder of the chapter focuses on Ivory and May, two of only seven girls in Carambo's chemistry class, which had twenty-two students. These girls also were student researchers in our science education research group over several school years and summers.

Othermothers

At City High, on average about two hundred girls are teen mothers (Lytle, 1998). Although none of the girls in Carambo's chemistry class were biological mothers, most of them, including May and Ivory, had child care responsibilities; they were "othermothers" or women who are caregivers to nonbiological children (Collins, 1990). The role of othermother is connected to the strong role of women and women-centered family networks within the African American community. Sociologists argue that the expanded role of othermothers demonstrates an ethic of care and responsibility that extends past traditional family divisions, and into a concern for the community. It is demonstrated in a disposition for communalism that is present in both males and females (Boykin, 1986), though more pronounced in females. It appears that this caring reach persists in families, extended families, and within other African American groupings even in the absence of financial need. However for the girls at City High, being an othermother appears to be connected to their families' fiscal realities.

Stemming from generations of outright slavery, followed by continued oppression, discrimination, and lack of opportunity, African American families lag far behind white families in the accumulation of financial resources. This is particularly true for the girls at City High, since many are from families that live in poverty and often have little financial security. Female teens such as May are human resources to the family unit in that they can supplement hard-to-come-by economic or material resources. May is an othermother to her sister's children and cousins, and as such she has to juggle her responsibilities as a caregiver and as a student. Some of May's school absences are attributable to a need for her to babysit her young nieces, nephews, and/or cousins when they are ill or home from school. The adults in the extended family generally need to work, and do not have the option of remaining home to nurse sick children. Her family does not have the fiscal resources to hire a babysitter, nor is it a common practice for African American families to pay a stranger to provide care in lieu of a relative or close family friend.

May's role as an othermother seems to restrict her participation in school and interrupt her studies, and could mean that she will not acquire the education she needs to pursue her goal to become a pediatrician or physical therapist. This is made more problematic by the school's rigid policies on lateness and absenteeism. Thus it is not just the othermother role that creates problems for May's education, but the interaction of her repeated absences with the structure of the school. It is the tendency of schools to see such an out-of-school role as a negative, while a both/and approach would perhaps reconfigure schooling to find capital in this phenomenon.

The caring and raising of children and other othermother duties affect girls and boys differently. Typically, boys are not expected to babysit younger siblings, escort them to school, or remain at home to nurse ill children (Burgess, 1994), although at times they do. Girls can also be assigned or assume othermother roles in school. For example, when boys have missed classes, teachers commonly ask girls or girls may volunteer to share their notes or tutor boys to "catch up." Another strategy is for teachers to place "disruptive" boys next to "studious" girls, with the expectation that the girls' behavior will have a calming influence on the boys. When girls assume such othermother duties in class, it may be to the detriment of their own learning. However, at times girls may gain from these othermother practices. Their own learning and understanding, for example, of science, may be enhanced by being a peer tutor, and at times they may gain social capital and respect for their endeavors.

For African-American women, motherhood, either as a biological mother, family othermother, or community othermother is a powerful symbol (Collins, 1990). Acting as an othermother in school may demonstrate the girls' respect for their peers. When girls willingly act to teach other students, they project care and concern for members of their community. Further, they assist their peers to garner cultural capital within the class, thus affording them agency, which may in turn contribute to collective agency. Through these actions, girls show their concern for the community and may, in fact, improve the well-being of the group as a whole. The potential of such othermothering for the development of positive emotional energy and group solidarity (Collins, 2004) has yet to be explored.

Being an othermother has taught May and Ivory how to care for younger children and made them aware of the responsibilities associated with raising children. Although the girls report that they enjoy caring for their younger relatives, these caregiving experiences have also led May and Ivory to express a decision to either remain childless or have only one child, since they appear to see how childcare responsibilities could restrict their goals.

Working with Black Girls

African American girls, whose lifeworlds place them at the margins or outside the dominant school culture, may exhibit practices that are agentic. These may be in the form of outspokenness or "doing school" with a quiet persistence. Both strategies challenge the low educational expectations that teachers and administrators have for urban youth and provide the girls with a mechanism to attain parental approval (Fordham, 1996). In the black community, academic success is often viewed as more important for girls than boys by both their parents and the girls themselves (Weiler, 2000). African American mothers expect their daughters to get an education and to develop self-reliance, and become resourceful (Collins, 1990). These qualities can cause conflict for girls in school environments where the dominant white culture expects students, especially females, to be quiet, compliant, and helpful.

As Tobin discussed in chapter 2, the science class is an example of a field where culture is enacted. Within fields, participants compete for cultural, social, and symbolic capital. The extent to which participants' positions are central or peripheral is determined by the types of capital that are salient in the field and how much of that capital they possess. However, the accumulation of cultural capital (i.e., learning of science) impacts a participant's ability to enact his/her agency in the science classroom. Agency is viewed as the power to act and change structures in a given field. Hence, if participants do not learn science, they have little power to act in Carambo's chemistry class and cannot change the structure of that field to better suit their learning.

Teachers are a critical resource for students' science learning. Teachers have high cultural capital with regard to science knowledge and have the agency to structure the learning environment through their curricular choices, teaching approaches, and expectations of their students' ability. Other research has reported that teachers often have low expectations for black girls (Mirza, 1992). In Mirza's study, teachers complained about girls who showed their high levels of self-esteem by assertiveness in the classroom. Although the teachers' negative attitudes did not reduce the girls' self-esteem, the girls reported that they often did not seek assistance from teachers who disrespected them and/or they avoided classes taught by those teachers (Mirza, 1992). In contrast to teachers in other studies, Carambo engaged the "loud" black girls in his class, providing a safe public space for them to use their voices in learning science. He did not silence them or admonish them for practices such as calling out or overlapping speech that others may label as "unfeminine." Carambo's patterns of engagement with girls who were outgoing and "loud" provided a classroom structure that encouraged students' assertive behaviors, such as calling out answers to his questions, demanding his attention, or dominating other resources.

Queen Mother of Da Bridge

Ivory and May provide contrasting examples of how the accumulation of social and cultural capital impacts agency in the chemistry classroom. Ivory is known as "Queen Mother of Da Bridge." Da Bridge refers to an area in Philadelphia near Ivory's home, where mainly male youth congregate and play basketball. Ivory's acceptance into the masculine area of Da Bridge's basketball courts reflects her well-developed abilities for playing basketball, and her identity as a female who can negotiate this largely male domain. Through self-definition, African American women often resist society's strong cultural stereotypes based on race and gender, or combine aspects of them in novel ways. In order to develop self-definition, girls need safe spaces, such as female friendships, mother/daughter relationships, and formal groups (Collins, 1990). Ivory has a close relationship with her mother, who allows her to play basketball at the local courts and supports Ivory's participation on various basketball teams. City High's girls' basketball team is a formal group and another safe space for Ivory, who regards her basketball ability, combined with academics, as one possible pathway to leave Da Bridge. Ivory notes that the skills she has learned on the basketball court have also influenced her studies by changing her attitude toward school. In an interview about her future, conducted by May, Ivory observes that she needs to "keep her eyes on the prize" by focusing on her studies and basketball. She displays the same calm perseverance that she employs in her basketball practice in the science classroom.

Ivory attains social capital through basketball as well as her ability to rap. In particular, she can "pop fly" or freestyle, that is, spontaneously create rap. The assertiveness and voice that Ivory demonstrates on the basketball courts and while rapping are also seen in chemistry class. She is confident and comfortable using her voice to engage Carambo and often used outspokenness as a strategy to express her scientific ideas and test her knowledge (Hurtado, 1996). Ivory has the social capital to access the resources available in the chemistry class and thus uses her dispositions in ways that are agentic. Further, her verbal engagements with Carambo ensure that she learns chemistry. Throughout the year, we observed Ivory garner cultural capital by engaging in the discourse of science by answering Carambo's questions, asking her own questions, and receiving Carambo's attention.

Ivory's commitment, assertiveness, verbal acuity, high self-esteem and self-definition are dispositions that allow her to succeed in class. This is in part because of the structure Carambo has developed in his class and his success in engaging City High students in science. During the time he has taught at City High, we have seen Carambo align his teaching style with the dispositions of his students. For example, other teachers may view the actions of the "gangsta chemist" as dangerous, but Carambo sees the student's moves toward the camera as *playin* and an indication that the student and his peers remained focused and were engaged in the lab. Carambo appreciates and understands the levels of personal energy, outspokenness, and use of verbal arguments that his students bring to the class, and he adapts his teaching to these practices. This results in more synchronous interactions, which often are successful and produce positive emotional energy. In such a climate, Ivory is advantaged by being able to enact practices that are similar to those she enacts in other fields and those welcomed by Carambo for participation in science class.

Teachers may find that high energy, oral patterns of engagement that emphasize verve, movement, and rhythm (Boykin, 1986) challenge their white, middle-class constructions of femininity, in which women are quiet. This is not the case with Carambo who is Afro-Cuban and whose teaching tends to be highly energetic and characterized by bodily movement, verve, and oral expressiveness. His teaching resonates with Ivory's dispositions and thereby provides a structure that supports her agency. Whereas, in many classes, Ivory is prevented from participating in a "loud" manner, in Carambo's class, she can excel by participating in ways that are natural to her.

Another consequence of Carambo building social capital and developing an understanding of the cultural capital his students bring from their neighborhoods to his classroom is that he exhibits an "ethic of care." There are three aspects to an ethic of care: personal uniqueness, the role of emotions within dialogue, and developing a capacity for empathy (Collins, 1990). During his first year of teaching at City High, Carambo had time to understand how his students' social and cultural capital impacted their agency within the classroom and he began to change the class structure to enable more of his students, such as Ivory, to fully utilize their capital from outside of the classroom as a resource for learning science (by doing science).

Ol' Head

In contrast to Ivory's outgoing persona as "Queen Mother," and outspoken engagement with Carambo, May has dispositions that lead to her interacting in ways that are reflective, quiet, and low in energy, body movement, and verve. Her fellow student researchers affectionately call her an Ol' Head, that is, a wise person. Although she can be playful in her interactions with other youth out of school, at school she is usually quiet and attentive and unlikely to call out without being called on or endeavor to speak without first raising her hand. So, in a context of students like Ivory speaking when they wish, and doing so with high energy and verve, May finds that her opportunities to participate are limited by her dispositions to be quiet and not to initiate interactions in this field. The structure of the classroom, especially as it is shaped by the oral contributions of a high-energy teacher and a small number of high-energy students places students like May in a role of having to participate mainly by listening. Hence, in the oral interactions she is at a decided disadvantage, first because her dispositions are not to act that way, but also, because her infrequent attendance at school makes it hard to think through the chemistry developed in whole class verbal exchanges. Even though these exchanges are supported by summaries, keywords, diagrams, and equations on the chalk board, May's attendance record and her rate of completing homework assignments do not allow her to readily access and appropriate these structures.

May's quiet manner and sporadic attendance at school (related in part to her othermother duties) meant that she did not build social capital with Carambo, who did not regard her as a strong science student. It is interesting that Carambo's teaching was adaptive to the culture of students like Ivory but did not afford the agency of students like May, who relished one-on-one interactions and was highly successful in learning science in such conditions (Elmesky, 2001). May would have benefited from being involved in cogenerative dialogues in which her home circumstances would have been more visible to Carambo and her willingness to learn might have created a contradiction for him to resolve. In such circumstances it is probable that Carambo and other students could and would have adjusted their roles to increase the opportunities for May to learn science in ways that suited multiple dispositions.

Similar to her heroine, the studious Raheema, in *Flyy Girl* (Tyree, 1993), when she attends school, May chooses to focus on her studies. Though she has difficulty matching the energy of Ivory and Carambo and rarely gains his undivided attention, May is resourceful. She often accesses other human resources, such as student teachers or members of our research staff who are frequently present in the classroom, to assist her learning of science. In this regard, May is agentic. But her agency is ultimately truncated by classroom structures that afford the outspoken, outgoing, and high-energy dispositions of Ivory and several others like her.

May has high social capital and is not disrespected by her peers although her demeanor is reserved. Because the African American community values motherhood in a variety of forms, May gained social capital through her work with us on various aspects of our research during the school year and the summer. These and her othermother experiences have enhanced her wisdom. However, that capital did not transfer into an expansion of her agency in science class or allow her to create social capital with her teacher.

Agency of Girls in Science

The multiple schema of class, race, and gender impact African-American girls' lifeworlds and schooling experiences in a multitude of ways. Generally, African-American girls attend school more regularly than their male peers, and girls from low-income families expect that they will enter and remain in the workforce for most of their lives (Weiler, 2000). African American girls graduate from high school at higher rates than their male peers. Yet, their roles as othermothers can adversely impact their education, even as it can enhance their social capital within their communities. Ivory and May are young women who have grown up in a minority culture, in circumstances of economic hardship, and face a different set of challenges compared with their Latina, Euro-American sisters or African American girls whose lifeworlds are spatially situated outside of the inner city. Yet schooling practices routinely fail to recognize these differences, and instead proceed with extreme cultural bias.

Laird (2002) challenged the adults in girls' lifeworlds to support, care for, and provide girls with the tools they need to value themselves. She suggests that "befriending girls" is a form of praxis and a micro-political strategy that can enhance girls' agency by providing them access to material, human, and spiritual resources. Contrary to Laird, I suggest that many girls, such as May and Ivory, already possess many resources and value their feminine attributes. Important community and personal resources are demonstrated by the girls; whether being assertive and loud or othermothering outside of school, they are undervalued and often looked upon negatively by the practice of schooling.

Carambo has had more success connecting urban students to science than many other teachers. In the eventful classes of urban schools, it may be difficult for teachers to befriend girls that are as different as Ivory and May. How can a teacher reach across age, class, ethnic, and sometimes racial lines to create social networks with students? For this to occur all participants will have to develop new practices that allow successful interaction to flourish. Carambo's efforts to align the curriculum to students' interests provide an insight into why students engage and disengage with science. How can we assist teachers to structure their classes to afford all students, particularly girls, with many forms of cultural and social capital opportunities to utilize their agency in the science class? The girls from City High are resilient and fragile, boisterous and quiet, determined and hesitant, mothers and othermothers.

Editors' Perspectives

Scantlebury draws attention to the manner in which the agency of female students can be truncated by the structures of the fields in which they participate. In Carambo's science classroom there was evidence that Ivory benefited from his culturally adaptive ways of teaching whereas May's dispositions were not afforded to the same extent. A possible solution to problems of this sort might be to use small group and whole class cogenerative dialogues. If May and Ivory were to sit with Carambo and some of the coteachers from the research team, differences in participation, agency, and structure would have been raised, if not by Ivory and May, surely by researchers such as Scantlebury. Then participants would be aware of the problem and cogenerate resolutions to remove contradictions about outcomes. Hence, the styles of interaction between and within groups and individuals could have been adapted with the goal of increasing the chances of them being successful.

A larger issue of how participation in the school field interacts with the science education field is evident in the problems experienced by many females, especially May. School rules on attendance disadvantage May because she is obliged to be an othermother in the family field. To overcome this disadvantage, school rules pertaining to absence from school need to be reconsidered. Cogenerative dialogues might assist to cogenerate fresh structures that would allow multiple interests to be addressed within the particular contexts that apply. Cogenerative dialogues involving parents/guardians, teachers, school administrators, students, and school district administrators might result in solutions that would allow students like May to fulfill their family responsibilities and continue to learn through participation in schooling.

A research focus on male students has high priority because of the continuing crisis of low graduation rates from high school for urban male youth and high levels of incarceration for urban youth who fail to graduate from high school, especially African Americans and Hispanics. Male youth, such as Shakeem and Randy, student researchers who appear in several chapters in this book, have the intellectual resources to reach almost any horizon we can imagine. Yet, their trajectories are a continual roller coaster. Shakeem once summed it up in a presentation he made at an international conference in which he described the contradictions he experienced as a "Gangsta and a Gentleman." Just as May was often disadvantaged by the structure she experienced in chemistry so too are male youth like Shakeem. We regard it as imperative to undertake research on the agency-structure relationships that unfold for urban males such as Shakeem and to use cogenerative dialogues to create communication skills to interact successfully across the boundaries of age, gender, ethnicity, and class. In the absence of cogenerative dialogues it is difficult to see how youth can learn to effectively communicate with their teachers who are often culturally and socially other. Similarly, cogenerative dialogues are a field in which teachers can build culturally and socially adaptive capital. Hence, cogenerative dialogues are not only places where collective decisions can be reached about rules, roles, and responsibilities, but also where cultural production can occur with the potential to be applied agentically in different fields.

Scantlebury's study draws attention to the significance of research that explores patterns and contradictions at the meso level of experiencing social life, augmented by studies of interactions at the micro level. In relation to May and Ivory, just how do their patterns of accessing and appropriating resources differ at the micro level? Since each student experienced a different structural environment, how did teaching contribute to those differences and to what extents were shut down strategies salient? Since most interactions are unconscious at the micro level, it is of interest to identify patterns of coherence and associated contradictions that might provide insights into differences in participation and achievement within and between groups based on gender.

Tracey: Teaching is a second career for me. My undergraduate degree is in civil engineering; therefore, my background in science is not very strong. To gain teaching credentials, I enrolled in a graduate teacher certification program at a local university. After one semester of education courses, no science courses, and no student teaching experience, I was given internship certification and placed in a comprehensive high school in Philadelphia where I was expected to teach chemistry and physics. This meant that I was given my own roster and the fate of those children, at least their fate in terms of learning science, was placed in my unqualified hands. Despite my lack of training and content knowledge, I approached teaching enthusiastically, but my enthusiasm quickly waned as I faced a floating schedule (I did not have my own classroom), teaching in class-rooms with more students than seats, and a new cast of characters each day as new students appeared while others left for undisclosed reasons.

To maintain internship certification and to work toward full certification, I had to continue to take education courses in the evenings. At times I found the contrast between my evenings at the university and my days in the urban neigh-borhood high school unbearable. I could not reconcile the contradictions be-tween what I was being taught in my education courses and what was and was not working in my own classrooms. At the university, I was taught that rote learning was unproductive and yet students seemed to demand these kinds of ac-tivities, rewarding me for giving such tasks with docile, complicit behavior.

After teaching for several months, I reached a breaking point in my level of frustration, so I put my lesson plans aside and decided to try to hold a dialogue with my students about what we could do to improve the class. I was shocked by their responses. To my surprise, my students did not appreciate this attempt to share responsibility for the class. In my mind, I was attempting to treat them as colleagues and not subordinates. However, students responded that it was "my job" to control and discipline them. They told me I needed to "make" them do their work, that I should "kick people out" who were not cooperating, and that we should "read from the book and answer the questions at the end of the chap-ter." I was chastised for not giving "notes" to copy off the board and vocabulary words to define. Although their suggestions contradicted my beliefs about teach-ing, I tried to incorporate my students' feedback into my teaching. On a later date, as I was giving notes using the overhead projector, one of my students was quick to point out to me, "Miss O., see how quiet everyone is." I was rewarded for giving notes to copy with a quiet, orderly classroom.

During my first, formative year of teaching, I could have easily been lulled into developing this kind of teaching practice. Fortunately, my department chair supported me by sharing lab materials and lesson plans and by setting aside time for us to collaborate on planning units for our physics classes. However, she did not teach chemistry and my background as a civil engineer did little to prepare me for teaching chemistry.

New Tools

Cath: It sounds as though you were starting to use some innovative strategies in your physics classes but found teaching chemistry more of a challenge. Did something happen that provided you with the opportunity to develop a richer knowledge of chemistry and to learn about pedagogical tools that you could use in teaching chemistry?

Tracey: Fortunately, in my second year of teaching at Southeast High, I was accepted as a member of the first cohort in a new Masters of Chemistry Education (MCE) program at the University of Pennsylvania. This program, for practicing secondary science teachers interested in expanding their chemistry content knowledge and improving their teaching practice, required me to complete eight chemistry and two chemistry education courses over three consecutive summers and two academic years. Involvement in this program was the life preserver I needed to keep from going under in my teaching assignment. By providing an opportunity to study chemistry more intensively, the program helped me to become more confident in teaching chemistry. Prolonged interaction with fellow students in the program enabled me to build a community of science teachers/learners with other participants in the program. This helped to alleviate the feelings of isolation I was experiencing as a chemistry teacher. The chemistry education courses, as well as exposure to some of the methods used by our chemistry instructors, provided me with pedagogical tools to use in my own classroom.

In the first MCE chemistry education course, we learned about many teaching tools. One particular tool called *levels of representation* (Gabel, 1999) refers to the practice of considering chemistry on three levels, namely: the macroscopic (what we are able to observe), the submicroscopic (what is happening on the atomic or molecular level), and the symbolic (how we communicate chemical knowledge, the language of chemistry). Understanding any chemistry concept involves being able to understand phenomena on all three of these levels and to move freely between levels for a given concept. As a successful learner of chemistry, this is something I have learned to do over time, and this practice has become a habitual, largely unconscious part of my chemistry learning. However, for high school students with less experience in chemistry, the need to think about and make connections between these three levels is not apparent. Learning about *levels of representation* in my university course made me conscious of this aspect of my chemistry learning habitus (Bourdieu, 1992), and this awareness gave me a tool that I hoped would benefit the learning of my students. High school chemistry, with its focus on symbolism, is often presented in a manner that treats these three levels as separate, disconnected entities. Thus, I was excited by the prospect of designing chemistry curricula using this tool as a way of helping students develop the habit of considering chemistry topics on all three levels and trying to make connections between those levels.

Cath: Why did you think this particular tool would be effective for your students?

Tracey: Perhaps because I was a new teacher, classroom management was not my strength. Earlier, I mentioned that my students would be very cooperative and well behaved when they were given tasks that involved rote learning, such as copying definitions of vocabulary words from the text. That, however, was not the only time students would be cooperative without coercion. Laboratory activities were the other type of activity in which students would willingly participate. Noticing this pattern, I began to wonder what were the features of these two very different types of activities that made students more willing to engage in them.

The demand from students for rote learning tasks, which I found very troubling as a new teacher, began to make sense to me as I started to explore the literature on urban education and as I became more familiar with the practices of other teachers and administrators in my school setting. Haberman (1991) writes of a "pedagogy of poverty" which refers to typical teaching practices that dominate urban settings consisting of actions such as giving information, monitoring seatwork, and punishing noncompliance. The majority of students in my chemistry class had a history of experiencing these teacher-centered actions as schooling. Therefore, these practices, aimed at controlling student behavior, were what my students identified as teaching, and what they looked for in those who taught them. Haberman argues that in many urban classrooms there exists a dichotomy where teachers teach and students learn but there is almost no interrelationship between these two sets of actions. The rote learning tasks associated with this type of teaching were familiar and comfortable for my students, hence their willingness to be compliant in performing these types of activities.

Unlike rote learning tasks, student compliance in conducting labs had little to do with comfort or familiarity. In fact, it was the opposite—getting a chance to do something they had not done before. Students willingly engaged in labs because they actually *wanted* to do the labs. They wanted to touch the equipment and manipulate the materials. The chemistry topic that was the focus of a specific lab might not have interested the students but they found doing the lab activity interesting, which motivated them to willingly participate.

My beliefs about the nature of science as a dynamic field where new discoveries and theories are constantly replacing old paradigms made it impossible for me to teach science as a set of facts. I wanted to expose students to the process of scientific experimentation and discovery, which I thought would be more enduring and useful to them than any specific science knowledge. I also knew that my survival in an urban school would depend on whether or not I could manage my classroom, and my initial attempts at an inquiry-based curriculum were classroom management disasters. However, as I learned to design activities drawing on students' interests (Seiler, 2001) and the knowledge they brought with them to the classroom (Elmesky, 2003), I found that "engaging" learning activities were my best classroom management tool.

Although students were eager to participate while conducting labs and they enjoyed and were successful at making macro-level observations, their levels of engagement dropped off dramatically when it came time to discuss what was learned about chemistry during the lab. In fact, students seemed to have little interest in making connections between the laboratory activities and chemistry concepts they were studying in class. I hoped that using *levels of representation* would help me to focus students' excitement for the macroscopic level, to ask why questions, to use molecular explanations for the macroscopic behavior of materials they observed, and to communicate those explanations symbolically using the language of chemistry. I also hoped that *levels of representation* would help students to see that lab and class work were not separate entities, but rather, that the lab was the physical manifestation of what we discussed in class and of what they read in their textbooks.

Cath: The idea that chemistry should be experienced macroscopically, microscopically, and symbolically can be traced as far back as Lavoisier, but only in the past ten to twenty years has it become a tool used to mediate the learning of chemistry (Wertsch, 1998). As a relatively new tool in chemistry education, there exists a high level of flexibility and ambiguity about its use for mediating the learning of chemistry, which, I think, is something that you experienced.

Other Resources

Tracey: Yes, teacher education programs can provide teaching tools, but this knowledge does not directly translate into being able to use the tool effectively in a different context such as one's own classroom. However, I need to emphasize that the chemistry education course also provided me with other resources such as laboratory activities that I used in conjunction with this pedagogical tool. After doing specific investigations in the chemistry education course at the university, we were asked to consider how each lab would fit into a high school chemistry curriculum and how we could use it in our own teaching. One such lab involved the making of "slime" from the mixing of a polyvinyl alcohol solution and a borax solution. In writing about how I might use this lab in my classroom, it occurred to me that it would lend itself to using *levels of representation* to organize instruction about polymers and cross-linking of polymers. Toward the end of the 2000-2001 school year (the same year I was taking the chemistry education class), I decided to teach a short unit on polymers, using *levels of representation* to plan and design learning activities where students would learn about polymers on each of the three levels. However, although I used *levels of representation* to develop lessons to teach polymers, I never overtly introduced my students to the concept of *levels of representation*.

The polymer unit began with an activity in which students used paperclips to model what polymers were like at the molecular level (SEPUP, 1992). Each paperclip represented a monomer, and students built long polymer chains. Placing their polymer chains in one cup and monomers (single paperclips) in another, students attempted to stir and pour the contents of each cup. They made observations noting the relative ease with which one could stir and pour their paperclip models of polymers and monomers. Using different colored paperclips, students linked their polymer chains together at different points on the chain (not end to end) in order to model cross-linking. Again, they compared and contrasted the stirring and pouring of the paperclip models of cross-linked versus unlinked polymer chains. They also built models of the polymers using molecule kits to explore how these molecules might be represented symbolically. In total, the class made polymer models for about three to four class periods (one hundred fifty to two hundred minutes).

Finally as the grand finale to the unit, the students made "slime," a gel in which cross-linking of polyvinyl alcohol (PVA) was initiated using borax solution (Orna et al., 1998). During the making of slime, students ran the same tests (stirring and pouring) on the PVA solution and the slime as they had with the paperclip models of polymers. Making slime took less than one class period (roughly thirty to forty minutes). Students were thrilled with the slime. They were able to observe the differences in properties before and after cross-linking of the polymer. I thought everything had gone well; after all, students were able to learn about the cross-linking of polymers on all three levels. As part of an assessment instrument I asked what I thought to be a very simple question, "What was the cross-linked polymer we made in class?" I was expecting students to tell me about the slime that they had made. Instead, virtually every student gave the same answer: paperclips. I was horrified. I asked myself, "What went wrong?"

What Went Wrong?

Cath: I can imagine your concern at this stage especially since you thought the question was so simple and the answer so obvious. Experiencing such an unexpected response from students can be daunting for teachers because usually they ask questions based on what they believe they have taught, and therefore, what they believe students should know. How did you make sense of this experience?

Tracey: I was convinced there was nothing inherently wrong with any of the individual activities we had used or *levels of representation* as a pedagogical tool. I realized that using this tool most effectively would require adaptation to my classroom setting. Wertsch (1998) calls this appropriation, and despite what appeared to be an initial failure, I was able to accept the setback as part of the process of appropriating tools from a graduate level chemistry education course for use in teaching high school chemistry. According to Wertsch, mastery and appropriation of cultural tools are separate actions. You can master the tool and use it in specific contexts, but appropriation involves "taking something that belongs to others and making it one's own" (53) or using it in novel contexts and modifying it as needed. Bakhtin (1981) argues that because tools have a cultural history, they resist our will and our appropriation, and that certainly seemed to be the case for me with respect to adapting *levels of representation* to my classroom teaching.

Using Design Experiments

Cath: How did you seek to overcome the "resistance" of this tool to your use of it in teaching high school chemistry?

Tracey: I decided a more structured approach to employing this tool might provide me with a better idea of how I could use *levels of representation* more effectively. I found Brown's design experiments (1992) helpful since the framework involved an iterative process of experimentation in which research informed my teaching practice, and my practice, in turn, raised new questions for research. This cyclic process of experimentation, reflection, diagnosis, modification, and then back to experimentation, encouraged me to think of the outcomes not as ends in themselves, but as the source of questions for new experiments. This framework for research was aligned with the MCE's emphasis on inquiry as a nonlinear and often cyclical process. Thus, when the outcomes were less than desirable, I did not see it as an endpoint, but as an invitation to further inquiry. This made me realize that the outcome I observed during my initial use of *levels of representation* was an invitation to further classroom design experiments.

I also realized that if I wanted to use a design experiment to examine whether use of macroscopic, submicroscopic, and symbolic experiences could make a difference to the chemistry learning of the students, I needed to begin by reflecting more broadly on my experiences and observations of the students' learning during the polymer unit.

Experience with *Levels of Representation*

Tracey: On reflection, I recognized that the amount of time students worked on building models of polymers (the submicroscopic level) was four to five times longer than the time they were involved in making slime, the real cross-linked polymer (the macroscopic level). This may have resulted in the paperclips being much more prominent in the minds of my students. Hence, the unanticipated responses to my assessment question led me to think about the role of the macroscopic level in the teaching and learning of chemistry.

Student interest is crucial to creating an entry point for students to become engaged with, and gain access to, science (Seiler, Tobin, and Sokolic, 2001). The macroscopic level is the most concrete, and it is the only level with which the students can make direct connections to the world they experience, making it a viable "entry point" for students to science. In addition, observations at the macroscopic level can serve as motivation for trying to explain what is happening on the molecular level. Johnstone (1982) describes how macro-level chemistry provides opportunities for raising scientific questions, designing experiments to test those questions, and then trying to interpret the results of the experiments to build scientific understandings. Moreover, he argues that chemistry literacy does not require students to have a detailed understanding of the micro level, that is, the level at which we conceptualize atoms and molecules. This argument about student learning is at odds with the commonly held belief that students should have a theoretical understanding of the science involved before engaging in lab activities, in other words, submicroscopic understanding before macroscopic experience (Chiappetta, 2001). Hence, it seems more natural for students to move from the concrete to the abstract rather than the reverse; and accordingly then, it makes little sense for me to have asked students to build paperclip models of a cross-linked polymer if they had not had experiences with substances that are cross-linked polymers. Rather, it seems more effective to begin a new topic or concept with an activity in which students can make observations and experience phenomena at the macroscopic level before they are asked to explain what is happening on the molecular level. Consequently, I wanted to design activity sequences in which students' first encounter with a particular topic was on the macroscopic level, especially since many of my students had not had many opportunities to work with materials in doing science.

In addition to analyzing the polymer unit, I began considering the teaching and learning activities that preceded it. I realized that I was expecting my students to glean for themselves the connections between the macroscopic and the molecular, that is, between the slime and the paperclips, when I had not overtly discussed *levels of representation* or used this way of thinking about chemistry prior to the polymer unit. In retrospect, it seemed obvious to me that in order for students to begin to move freely between *levels of representation*, this framework would need to be used consistently throughout the year, permeating all topics, not just polymers. Based on this awareness, I decided that the following school year I would emphasize, from the beginning, the need for students to understand chemistry on all three levels and, where possible, weave all three levels into each topic in my teaching of chemistry. In terms of the polymer unit, I planned many changes especially regarding emphasis and sequence of each of the levels. Rather than spending one class on the macroscopic and four on the submicroscopic level, I reversed the emphasis. Also, instead of beginning with modeling what was happening on the molecular level, students began with activities in which they could observe physical properties of polymers and cross-linked polymers before delving into the paperclip models. Several new activities were added to give students a greater variety of macroscopic experiences in which they could "get their hands dirty" and experience polymers.

Learning about Polymers: Revisited

Cath: In this design experiment, how did you change the sequencing and emphasis of *levels of representation* in your revised polymer unit?

Tracey: The polymer unit was carried out within the space of one week, which was similar to the previous year. On the first day, students synthesized a cross-linked polymer by combining Elmer's glue, water, and borax and explored its properties—a macroscopic experience with polymers. On the second day, students gained awareness that cross-linked polymers are found in common household products by dissecting a diaper. Students went to the lab on the third day to make slime, and it was not until the fourth day that we moved away from macroscopic activities and began considering what was happening on the submicroscopic level through the modeling activity using paperclips.

Cath: Could you briefly describe each activity?

Tracey: The first polymer lesson involved students in a role-play activity in which they worked in groups pretending to be research chemists trying to improve the design of a product, a bouncy "gluep" ball. As a CEO of a company, I had the rights to a gluep recipe with a 1:1:1 ratio of Elmer's glue, borax, and water. The resulting gluep was far too runny to make a bouncy ball. The research groups were given a budget and costs of materials, and they were charged with the task of optimizing the formula for gluep to make the bounciest ball (Opper and Spenser, 1997). In the process of making their ball, students began to observe how increasing the proportion of one material changed the properties of their gluep. Even students who rarely participated in classroom exercises became actively involved in this activity. At the end of the class period, the products were tested to see which research group came up with the bounciest ball. The students became rather competitive in their pursuit of the best gluep recipe and, for some, all thoughts of a "fair test" were ignored as they became absorbed in their efforts to gain the respect associated with having the ball that bounced highest.

The following day, in order to look at a familiar, everyday application of a cross-linked polymer, students dissected a disposable diaper to extract the polymer (Geuther and Olmstead, 1996). Polymer extraction required some patience and a bit of tinkering with the technique. However, once students had extracted as much of the polymer as they could, they measured the mass of the dry polymer and added measured amounts of water to determine the maximum amount of water the polymer could absorb. When the polymer was saturated, they reweighed the wet polymer to determine the mass of water absorbed. Students were amazed by the ability of small amounts of this polymer to absorb large quantities of water. They were also impressed by the fact that once absorbed, no amount of pressure could release the water from the structure of the polymer. One student remarked that now he understood why babies' diapers were so "saggy." Finally, using two diapers, they added measured amounts of water to one diaper to see how much water the diaper could absorb. In the second diaper, they added a 1 percent salt solution (simulating urine) until it reached saturation. Students indicated their surprise at the small quantity of urine relative to the amount of regular water a diaper could hold.

On the day following the diaper dissection, we went to the chemistry lab to do the slime activity (Orna et al., 1998), modified slightly from the previous year. Students made observations about the relative viscosities of water, polyvinyl alcohol (PVA) solution, and eventually slime. After making the slime, students were asked to play with it and make observations about the properties of the slime and how it responded to specific activities such as rolling it, trying to bounce it, and even leaving it sit on the table undisturbed for several minutes. Unlike the previous year, students spent more time manipulating and making observations of the slime.

Finally, after being exposed to many macroscopic experiences with cross-linked polymers, students made paperclip models. This was the same activity used the previous year where students connected paperclips to make polymer chains and then connected the chains with other paperclips to cross-link their polymers. The paperclip activity concluded the polymer unit.

Cath: Having made these changes, did you see a difference from the previous year in the assessment?

Tracey: If anything, I noted that students were more engaged this year than they had been during the previous year when I introduced polymers. In order to evaluate whether the change in emphasis from microscopic to macroscopic activities had made a difference, I decided to include the same question in their assessment: "What was the cross-linked polymer we made in class?" I was very hopeful as I reviewed their responses, but I was crestfallen to discover that a substantial portion of the class insisted that the cross-linked polymer we made was paperclips. The number of "paperclip" responses was reduced from the previous year, but overall the class's responses indicated that this was still a point of confusion. My initial reaction was: "I give up!"

From Sequencing to Integration

Cath: I was surprised when you told me the outcome of the question and I know that you were also. Now you have had some time to think about your actions and those of the students, what has your analysis of this design experiment suggested?

Tracey: Initially, as you suggested, I questioned myself about possible explanations for the students' responses to my question. Issues of timing and pacing were certainly significant. This unit was completed just before beginning review for final exams, and some students were openly resistant to learning new material. Also, because I wanted students to experience all the activities, we rushed through each activity never quite finishing and never pausing for closure for any of the polymer lessons.

Although an obvious hindrance to student learning, blaming failure of the unit solely on timing would be missing an opportunity for further appropriation of this tool. When considering my use of *levels of representation* in planning and implementing the polymer activities, I believe I fell into the trap of becoming overly concerned with the sequencing and separation when perhaps I should have been emphasizing integration of the levels. Students need to recognize the characteristics of each level and be able to use them so that they become part of each student's way of acting and being in the chemistry classroom. But by compartmentalizing macroscopic, submicroscopic, and symbolic into separate activities rather than incorporating each level into each activity, the connections between levels were obscured rather than highlighted. Students needed to see that the levels are merely three ways of representing and understanding the same thing, not three separate things. The type of modification I am proposing is that for any concept, the modeling or explanatory activity should be integrated with the macroscopic activity. In this way, students model the phenomenon that is before them. For example, rather than worrying about which activity students should do first, make paperclip models or slime, I would have them do these activities concurrently. Students could get their PVA, and then make paperclip chains to model the structure of molecules of PVA. They could then add the borax solution to cross-link the PVA and, immediately following making the slime, they could cross-link their paperclip chains. In this way, they would be able to make comparative observations, with both their paperclips and their slime side by side on the lab bench.

Making Levels of Representation My Own

Cath: As you thought about the variables that you modified and the assessment item you focused on, what did you learn from this experience?

Tracey: It is obvious to me that I sank into the pitfall of focusing on students' responses to one question, and not a very good question at that, to judge the success or failure of the polymer unit and my use of *levels of representation* in facilitating student learning. Seiler, Tobin, and Sokolic (2001) write about a similar loss of peripheral vision:

> We went into the activity sequence with the goal of using it as a vehicle to show the relevance of the laws and equations of motion and to create deeper understandings of Newtonian physics. Our focus on specific science distracted us from a potentially promising approach that centered on what the students were saying and doing. We now see the potential of ascertaining accurately what students are saying and doing in the class and teaching them accordingly. Also we are cognizant of potential dangers of teaching too narrowly toward goals we have identified as desirable prior to a given lesson. (18)

Broadening my vision and reviewing videotapes of the polymer lessons, I observed students who rarely participated in class actively engaged in the activities. Students who had previously shown little interest in science were using science skills to experiment with different proportions of glue, water, and borax to make the bounciest gluep ball. Students who often did not participate in lab activities were genuinely excited about making their own slime. During the diaper dissection, students were carefully measuring volumes of water and the mass of super absorbent polymer (SAP) they had extracted from their diaper. The investigation and dissection of diapers led many students to ask questions about how something so small could absorb so much water, and why it could not absorb as much urine as water. Student excitement and interest, student-generated questions, and students engaging in scientific behaviors such as experimenting and modifying experiments to answer their own questions were not measured by the simple question upon which I chose to focus. As I was bemoaning my failure to get students to see that paperclips were not polymers, I overlooked the scientific practices that were occurring around me.

In addition to the limitations of the assessment question, my commitment to the structure of the design experiment also blinded me to opportunities for further investigation suggested by student questions. In retrospect, I see that I became so concerned with students getting to do all the planned activities that I neglected the important component of communication and responsiveness. Our tight schedule left little time for students to discuss what they were observing, and to begin to refine their thoughts about possible explanations for the observed phenomena through writing. Engaging students in verbalizing their observations, questions and explanations is a critical component of the symbolic portion of *levels of representation* and of helping students develop scientific dispositions.

Cath: Using the videotapes we recorded, can you cite evidence of students engaging in scientific practices or increases in student interest?

Tracey: Students who rarely participated were highly engaged during the polymer unit. I also found that students who always do their work, but rarely do it enthusiastically, became much more excited during the polymer activities. For example, while reviewing videotapes from the polymer unit, I noticed that three studious students, Tran, Ju, and Lucy, were more animated than they had been all year. Although they were always diligent and worked carefully when they were engaged in an activity and they achieved high standards, during the polymer activities they were genuinely excited by the observations they were making. In one particular segment, Ju used a graduated cylinder to carefully add one hundred twenty milliliters of water to the SAP that they have collected in a four hundred milliliter beaker. Then Tran held the beaker upside down over his head and no water ran out. It all seemed to have been absorbed into the polymer.

Tran	That's so amazing!
Cath	Do you know how polymers form from monomers?
Tran	Monomers join to form polymers. (*gestures with his hands*)

Cath: Overall, did you feel this design experiment was successful?

Tracy: Since students' responses to the cross-linked polymer question were not what I expected, I was forced to consider what I was really hoping to learn through conducting this design experiment. Was I looking for the "right" way to use *levels of representation* to teach urban high school chemistry students? My initial revision of the polymer unit seemed to have been motivated by a quest to determine the correct way to sequence the levels and the optimal ratio of time to spend on each level in order to maximize students' understandings of polymers. However, as the experiment progressed, and especially when the results did not meet my expectations, I began to examine more subtle aspects of the adaptation of this tool to my particular classroom. The unanticipated results of the design experiment forced me to stop looking for the right formula for using *levels of representation*, and redirected my focus on the process of appropriation of the tool for both me and my students.

Beyond Simple Solutions

Cath: You have talked extensively about *levels of representation* as a pedagogical tool and you have cautioned against adopting a simplistic proposal about what "works" in teaching and learning science in an urban classroom. Are there some more general comments that you wish to make that you think might assist other teachers as they use specific pedagogical tools in their classroom and as they try to establish a context in which students also begin to use these tools?

Tracey: *Levels of representation* is a pedagogical tool that I adopted to mediate the learning of chemistry in my class. If the design experiment indicates anything, it indicates that to understand how this tool is used and to evaluate its effectiveness requires allowing sufficient time for its usage to develop in the science classroom (Cole and Engeström, 1993). Focusing on just one short unit of the chemistry curriculum does not provide students with sufficient opportunities to learn to use *levels of representation* as a tool for their own learning of chemistry.

Also, as a tool, *levels of representation* is relatively new and this has implications for structuring classroom interactions. Polman and Pea (2001) highlight this conundrum in their analysis of the use of inquiry in science classrooms where there does not exist a well-established set of rules for structuring interactions. The result can be that students experience contradictions as they endeavor to achieve their objectives and are unsure of teacher expectations.

If pedagogical tools such as the one explored here are to change the way students think about and engage in chemistry, there needs to be an ongoing commitment to using the tool in the classroom. Appropriation of a pedagogical tool requires a willingness to experiment with it in the context in which the tool is to be used. Appropriation requires that students and teacher have a certain level of agency to use the tool and an awareness of the values and meanings embedded in the tool (Wertsch, 1998). Appropriation also assumes that students will adopt the tool as they come to recognize its value in helping them to achieve objectives they have for learning chemistry and transform the tool to fit their own needs. I believe this is one area of student learning I could have examined more closely, for this transformation is based on students' negotiation of the use of this tool within the specific contexts in which they are working.

As students and teacher use a tool repeatedly, their use of the tool becomes part of their habitus for learning and teaching, part of a science teacher's or science student's way of acting in the classroom. However, achieving this level of appropriation requires intensive and consistent employment of the tool to refine rules for its use so that the tool can assist each participant to achieve his or her objectives for learning science. In my search for tools that enable the students in my urban classroom to make sense of chemistry, *levels of representation* offers a methodology that can assist students to recognize the importance of three levels that constitute the world of chemistry and to make connections between them. When used thoughtfully, it can strengthen both the students' experiences with materials and the connections they make between chemistry and their lifeworlds, two important goals in teaching minority students who have historically been marginalized from science learning.

Chapter 13

An Autobiographical Approach to Becoming a Science Teacher in an Urban High School

Sonya N. Martin

It took only one phone call home after officially entering the Graduate School of Education at the University of Pennsylvania to have me seriously examine my reasons for becoming a teacher. "You don't even like kids," said my incredulous grandmother on the other end of the line. After our conversation, I began to think about the audience for my future teaching career. I realized that, in fact, I had very little contact with children other than my peers in my own childhood. When asked now, at the end of my fifth year as a teacher, "Why did you want to become a teacher?" I respond, "I didn't set out to *become* a teacher, it just happened."

On the Outside Looking In

Conflicting Views

Being raised in a household with conflicted values concerning education, I was uncertain how to feel about school and my teachers. My grandfather, who by the age of fifteen had not yet advanced past the fifth grade, insisted that I "learn my letters" and "learn how to do arithmetic so no one could cheat me out of my money." As a sharecropper in the rural South, he was allowed to attend school only a few months of every year, and, as a result, he was never able to advance. He stayed in school longer than many of his ten siblings, but adult responsibilities and shame finally forced him to quit. My grandfather often lamented the fact that he was unable to "get an education" because he felt it was the only way to get a "good paying job." Each afternoon, upon returning from school, my grandmother made sure that I "got my lessons up" before going out to play. She sat next to me as I completed my assignments, assuring me that even though she could not help me with my work, she would sit next to me until I was finished.

As a child, no one ever told me, "You can be whatever you want when you grow up." I was never told what I *could* become; only what I should *not* become. I was forbidden to ever work in a sewing factory or a chicken plant, jobs held by most of the women in my family. My mother, who gave birth to me at age fifteen, told me on a daily basis that I had to do my best in school so I would not end up like her or my grandparents. Her words insinuated that she and my grandparents were not acceptable and had failed in some way. Their persistence that I succeed in school drove me to achieve academically as I knew they were all counting on me to do what they had been unable to accomplish themselves. I was made aware of the importance of "getting an education," but neither my grandparents nor I really knew what that meant, or the cost we would pay to achieve our goal. In fact, growing up, the cultures of my home and school were often at odds, both academically and behaviorally. While being urged to excel in school, I was chastised and called lazy when found reading a book for leisure, something for which I was praised at school. My home culture often clashed with school expectations of behavior as well. In second grade, a bully began to harass me and when the teacher did nothing to stop him, I told my grandfather. To my surprise, he spanked me and said he would do so every day until I stopped the bullying by "showing respect for myself." The next day when the boy pinched me, I picked up a chair and struck him in the head, sending him to the hospital for stitches. Although disciplined at school, my grandfather rewarded me at home, proud that I had taken up for myself, assured that "no one would mess with me again."

She Has So Much Potential

By high school, I had achieved autonomy from my grandparents in terms of school-related issues. Administrative needs often seemed foreign and alienating to them as they had difficulty reading and felt intimidated by status and position.

For this reason, teachers often consulted me directly to discuss my academic performance, and they rarely included my family members as part of my educational experience. Teachers often commented about my *potential* as a student and my *potential* to succeed and the *potential* to be something more than my family members had accomplished. By "potential" I understood them to mean that I had talent without opportunity. So, although I was driven to excel in high school, I never considered what would come afterwards since I had no concept of the opportunities available to me. My life experiences, at that point in time, did not lead me to expect that I would go on to college. Moreover, I had no intention of applying to college and made no plans to do so. However, in my junior year, a counselor arranged for me to take an AP exam in U.S. history, paying the cost himself. The scores from the exam were reported to various colleges, and shortly thereafter I began to receive information from their recruitment offices. The counselor insisted that I apply to at least one college, so I randomly chose and applied to Bryn Mawr College, a small women's college in Pennsylvania, making a decision that would alter my life forever.

Act Like a Lady

When I was asked to travel to Atlanta, over two hours away, to interview with the regional college representative, both the school administration and my counselor were thrilled at my success. A woman from the Board of Education took me shopping to buy a new outfit that would be appropriate for the interview. I was asked to attend local meetings of the Daughters of the American Revolution where I was taught how to "act like a lady." These women drilled me on how to properly cross my legs, how to speak, and even how to properly drink tea—all so I could conduct myself appropriately during the interview! The school also arranged for me to be driven to Atlanta. Upon acceptance to Bryn Mawr College, I was struck by the reaction of the school and my teachers. While they certainly had a hand in my acceptance, they completely neglected the contributions and sacrifices my family had made to help secure my success as well. Their reaction also seemed to negate my own role in the college's offer, making the achievement seem impersonal and unconnected to me.

The Great Divide

On scholarship to an elite private college whose traditions and customs were outside my realm of experience, I often felt out of place and isolated. I struggled to make sense of my new environment yet received constant reminders that I did not belong. I began to be positioned as "Other." Students and professors alike continually made fun of my heavy Southern accent, suggesting that my slow manner of speaking was directly equated with my intelligence, or lack thereof. This bigotry confused me as I was surrounded by people of races, religions, languages, and accents unlike any I had ever experienced before, yet no one made fun of them. Jokes such as, "If your mother and father get a divorce, will they still be brother and sister?," were commonplace, acting to silence and separate me from my peers. During visits home, I experienced the same ridicule in reverse from friends and family, who accused me of "putting on airs" as I unconsciously began to lose my accent and adopt new mannerisms. Surprisingly, my family's desire for me to succeed academically and to *become somebody*, eventually distanced me from my family, both physically and emotionally, by positioning me again as "Other," this time in comparison to them. I struggled with this double-sided alienation, both socially and academically, during my first two years of college as I attempted to hang on to my former identity while trying to fit into a new environment. While that was a difficult period of my life, I later realized that my experiences before and during college greatly influenced the type of teacher I later became and the types and quality of relationships I developed with my students.

Settling on Being a Teacher

Uncertain Certainties

By my sophomore year, I was firmly established as a pre-med biology student with a concentration in neuroscience. I chose to major in biology during my first year partly because I enjoyed the subject but mainly because it was the only subject that was not novel to me and in which I seemed to succeed. Similar to my high school experience, I had no template for how to proceed in college, having had no one after which to pattern my behaviors. For this reason, I relied heavily on the advice of my dean and professors to make decisions. In the beginning of my senior year, I began the process of applying to different medical schools and preparing for life after college. However, I still held a very limited view of what a person could do or become that would be considered successful. I believed becoming a doctor to be the pinnacle of success, so attending medical school never really seemed to be a choice at all, only what was expected from me—living up to my potential. Consequently, I had not really considered what it might take to become a doctor, only that my family, friends, and professors were happy with my decision to proceed in that direction.

A Life in Transition

An early morning phone call in the second semester of my senior year altered my plans. My mother, my two-year-old sister, and my best friend from high school were involved in a serious car accident. My experiences with my mother in the intensive care unit led me to withdraw my medical school applications. I decided I could not perform the duties required of the nurses and physicians who had tended to my mother and family, including helping us make the difficult decision that finally ended her suffering.

After the funeral, I immediately returned to school to finish my coursework, as my scholarship would not extend past May. The next few months were a blur as I tried to concentrate on completing my senior thesis and passing final exams. After graduation, I began searching for an alternative to my long-standing plans for medical school. I saw no future in returning home, so with the help of an alumnae connection, I secured a research position at the Veterinary Hospital at the University of Pennsylvania. The benefits of the position allowed employees to take courses at the University for free. Feeling that I would be at a loss without the structure of school and an academic learning environment, I gladly took advantage of the benefit.

Developing Interests

While working at the University of Pennsylvania, I enrolled in a human development course that examined both the physiological and psychological development of humans, but that mostly concentrated on childhood development and the process of learning. Many of my classmates were undergraduate education majors and they suggested I continue with them in the spring to the next course, anthropology of education. It was during that course, nearly two years after my mother's death and graduation from college that I began to consider pursuing a degree in education. I began to re-examine the world around me and to re-evaluate my own educational experiences and life expectations. The readings for the course radically changed my life view and goals. Books like *Jocks and Burnouts: Social Categories and Identity in High School* (Eckert, 1989) and *Ain't No Makin' It: Aspirations and Attainment in a Low-Income Neighborhood* (MacLeod, 1995) awakened something inside of me I had not known was there. I became very interested in theories about culture and the powerful influence of social reproduction in people's lives.

I found that the theory of social reproduction was of particular significance to me. I was the first person in my family to graduate past the eighth grade, and, at the age of twenty-four, I was unmarried and childless, a feat in and of itself in the community where I was raised. What piqued my interest most about the theory was the glaring contradiction of my life experience to that of my family and most of my childhood friends. I was an anomaly because I had escaped their fate. And so it was through the words of people like Bruner and Dewey that I began to consider the purpose of education in society and its influence on a child's development and socialization. At this point my interest in educational theory was on a personal and academic level as I had not yet begun to consider the practical implications of educational theory to teaching and learning.

The Next Step

By the following fall, after completing two more education courses, I decided to join the teacher education program full-time. It would be nearly two years before I encountered my first pupils. When I finally arrived at student teaching, the theoretical suddenly became practical as I met my morning and afternoon classes, each consisting of twenty eager kindergarten students. By the end of the day, my grandmother's words had played through my head a hundred times, "But you don't even like kids."

The next four months were a mixed blessing of sweet, enthusiastic learners and a firsthand opportunity to experience many of the problems associated with education about which I had been reading. My placement in a white, middle-class, suburban setting was not what I had expected, but during that time, I learned a great deal about the kind of teacher I did and did not want to become. I went to a different school mid-year, completing my student teaching with two different cooperating teachers in an urban magnet school. My new placement in a fifth grade science class and a ninth grade biology class allowed me to become certified to teach grades kindergarten through twelve. Yet even more importantly, I reached the realization that it was not that I did not like children, but that I liked working with older children best.

Theory and Practice

It was during my time as a student teacher that I began to realize the possibilities of how I could teach both responsibly, by presenting my students with accurate and objective information, and responsively, by tailoring lessons to promote real thought and discussion in my students. As an educator, I realized I wanted to encourage students to think independently, to question knowledge, and to learn to relate the knowledge they acquire to their life experiences and vice versa. I became certain that students would feel empowered by the learning process when they developed confidence in their thinking abilities so as not to rely on others to decide what knowledge is valued or valuable for them to acquire. It became very important to me as a new teacher to help students foster ownership of knowledge and experience. Education, as I envisioned it, was about providing students with the tools to explore and think about the world around them, not about learning what it means to succeed as defined by someone else. My beliefs about education, although better developed by my coursework in the education program, actually grew from my own experiences as a student. Thus, my main educational goal as a teacher was to assist children to think and to become lifelong learners in order to develop their own sense of success based upon their own values and culture.

One of These Things Does Not Belong, or Does It?

Theory Out the Window

My first few hours of solo teaching in an urban setting, sans cooperating teacher, convinced me that I had no idea what I was doing and that the only thing of which I was certain was that I was a horrible teacher. Looking around the room at the thirty-five African American students staring expectantly at their new, inexperienced white teacher, they knew I would not last three days. And so it began—the daily struggle, tortures, and trials that became my first year of teaching at Dewey Middle School in West Philadelphia.

The students and I began each day anew, waiting for the unexpected and reacting the only ways we knew how. I realized early on that having seventh grade level books in a seventh grade class was not ideal if the majority of students read on a third or fourth grade level. My class was self-contained, meaning that I taught all subjects. I quickly learned that novel, interesting lessons would not be a mainstay in my classroom for I was overwhelmed by the needs of the students and my inability to gain control of their behavior. My class, it turned out, had been together for the past two years, in which time they had over thirty different substitutes and never a regular classroom teacher. It seemed as though everything I tried failed. Student behavior seemed wildly unpredictable. I tried very hard to be the type of teacher I thought I was supposed to be, but I met with failure again and again. I came to school each morning with an upset stomach and went to sleep each night crying about the events of the day. I felt alienated from my students, by race and by the beliefs we held about one another. But over time, I began to understand how our cultures, despite being different, were also similar in many ways.

Earning Respect and Gaining Acceptance

I quickly found that while my ideals had worked in an environment where structure had long been established, like at the school sites chosen for my student teaching experiences, the students in my class practically revolted if the periods were switched in which spelling and math were taught. Nontraditional teaching styles were not well received in my new environment but I slowly learned, through months of trial and error, to institute a sense of order and consistency in my teaching upon which my students relied. Although my class was far from perfect, the students began to accept that I was not going to abandon them, no matter how hard they tried to make me leave.

This realization occurred sometime in the second month of school when a transient student re-enrolled in the class. By that time, I was used to students leaving and re-enrolling as they and their parents moved in and out of our school boundaries. However, the announcement of the arrival of this student in particular sent my class into chaos. They began immediately to recount stories of how she had hit one of the substitutes in the past and how she had made several teachers depart, one even leaving during the middle of the day without informing the office. Many of the students were openly concerned that I would finally leave them, and I must admit, their stories did little to set me at ease. As it turned out, the "showdown" was as spectacular as they had envisioned. Within an hour of her arrival, Cieara began challenging me in front of the class. She refused to do assignments or participate in class activities. As her misbehavior escalated, I noticed the other students becoming increasingly agitated, in anticipation of her inevitable eruption. At one point, Cieara jumped from her desk knocking it to the floor screaming how she had hit that "white teacher in the fucking mouth" and now she would do it again. I had tried to maintain what I envisioned to be the actions of a mature, responsible teacher during her previous outbursts, but her threat angered me so that I snapped! In a grand sweeping motion, I knocked everything from my own desk while yelling, "If you think you want to hit me then come on!" I then added, for the benefit of the whole class that "my own mother had whipped my ass worse than they ever could," so if any of them wanted to try and hit me, they now had the chance! Much to my surprise, and relief, Cieara sat down and the class worked silently for rest of the period.

A Turning Point

While this was the most extravagant display of violence to grace our classroom, it afforded me a new sense of respect from my students and even from myself. To me, this incident was a reminder that I would sometimes need to take a situation into my own hands, even if it meant handling the situation in a way that may not be sanctioned by the dominant school culture. The attitude of my class changed dramatically after that day. I feel the students decided that because I had stood up for myself, I not only cared enough to stay, but I had the ability to "make it" in their environment. Students often commented that whenever I was upset, my accent became heavier, often eliciting a warning from one student to another to "lay off, 'cause Ms. Martin gettin' ready to go backwood on you." It was after this incident that I began to handle situations more instinctively, listening to the needs of my students in a new way, and I found that we had much common ground upon which to build a better relationship. I began to notice that students responded more when I dropped my guard and spoke with a heavier Southern accent. Students said it was comforting and reminded them of their grandmothers and other relatives who were either from the South or whom they had visited in the South. Many of these students' families had migrated to the North from Southern states and being poor and from the rural South meant that I shared many cultural experiences with them, including food and religion. I found that I was in a situation where these students were actually appreciating my culture and I found that I could appreciate theirs. So while I still felt like a failure as a *teacher* (someone that disseminates information and knowledge), I knew I was committed to my students and I was slowly building relationships with them and their families.

It was also during this first year of teaching that Ken Tobin, the head of the education program from which I had just graduated, proposed that he bring his class from a local high school to work with my class on science projects on a daily basis. I was horrified! Although I was gaining some ground with students in my class, every day some event would occur that would rock my foundation as a novice teacher, and now my mentor wanted to come and be witness to my shortcomings. I reluctantly agreed to his request, hoping to learn from his example and, thus, I began down the road to *becoming* a "real" teacher.

A "Real" Teacher

Well, this was it. I was going to see how a *real* teacher would handle these students. I was going to see magic in the classroom, the likes of which I did not possess. What I saw permanently changed my view of myself and of teaching. My mentor came with students from *Incentive* SLC at City High to *do science* in my classroom (see Tobin, chapter 2). The experiments themselves were not eventful, nor were the student experiences. What was important for me, as a teacher, was to see my mentor, a teacher of thirty-plus years, struggling to *teach* the students in my classroom. I had honestly thought an experienced teacher would work wonders with my students. I thought the students would respond with new enthusiasm and become instant scholars. What I saw was a man struggling to control the students and making many of the same mistakes I made daily. I saw an experienced professional failing "to teach," and my perception of *my* ability to teach my students changed dramatically.

Suddenly, I realized there was no magic to teaching. There was no easy way to reach my students. And although our science outings eventually ended with one of my students dropping a potted bean plant out of the third-floor window onto one of the high school students, I had a new sense of what could reasonably be accomplished. This transformation in thinking required losing my image of what a teacher was supposed to be in order to become the teacher my students needed. I began to look at each day as a new beginning and kept trying new methods. I began to handle situations more instinctively, listening to the needs of my students in new ways, and finding that we had many common experiences upon which to build better relationships. I learned the necessity of giving respect to students by valuing their culture and knowledge while simultaneously demanding respect from them. I worked hard to understand their vision of the world and to help them see mine. While growing up, I felt that teachers did not communicate to students on a personal basis, but at Dewey Middle School, I learned to relate to my students on a personal level and them with me. As the year advanced, I traveled to the police station on three different occasions for student-related crimes and I was present at the sentencing trials for two of those students. I attended three wakes in the homes of my students for lost family members. I visited students in shelters and hospitals, and I purchased clothes and toiletries for a student, without parental assistance, who was placed in a mental health facility. I held children while they cried and I held the hands of parents and grandparents who did not know what else to do for themselves or for their children.

Today, when I look back on my first year, I do not remember teaching my students much in the traditional sense of the word. Yes, we examined worms and sea monkeys in science class and traveled to the art gallery to study impressionist art. But the real lessons I taught and learned that year were associated with how to develop a community. I had class discussions in which boys and girls alike cried openly, expressing their fears and concerns for their future lives and those of their families. We discussed racism, sexism, poverty, and drugs and the roles of each in their lives. We built a community based on trust, where I learned that being a teacher is not only about books, tests, and projects, but also about meeting the needs of my students in all aspects of their lives. So while I often felt like a failure as a teacher in the traditional sense of the word, I came to a significant realization about the kind of teacher I wanted to be. I realized the importance of listening to my own voice as a person and the value of sharing my experiences with my students on a personal level as a means of building community and trust between teacher and student.

The Other Side of the Coin

Starting from Scratch

I still remember walking into my first chemistry class at Urban Magnet. I watched as the students filed into the room, choosing their seats for the term. Coming from my last teaching experience in an underachieving, predominantly African American school, I was challenged by the racial diversity represented by the class population. The class was about 60 percent white and 40 percent African American, Asian, or Latino. After a brief introduction, I asked the students if they had any questions. One student raised his hand, leaned back in his chair, and with his head cocked to one side asked, "Is it true that you're not even certified to teach chemistry?" I was completely unprepared for this student's question and, taken off-guard, I admitted to the class that although I was a science teacher, I had not been certified to teach chemistry. I could see I was quickly losing the confidence of these students as they began to bemoan the fact that they had been scheduled into my class.

A New Curriculum

As seen in my first encounter with the students at Urban Magnet, my new position held many surprises and challenges for which I was unprepared. Although certified to teach elementary education and secondary biology, I was assigned to teach a tenth grade academic chemistry course, two eighth grade physical science courses, and a ninth grade biology course. Needless to say, I was overwhelmed. I had always been enthusiastic about chemistry, but my content knowledge was severely lacking.

So now, in a new school, as a new teacher, I would be dealing with an unfamiliar curriculum and unfamiliar students. It did not take me long to understand that my first year of teaching in a "typical" urban school setting had prepared me to control students, but certainly not to teach them science. Each night I busied myself preparing lengthy scripts, including notes as to when to write on the board and even jokes to add if the classroom climate allowed. I practiced chemistry problems each night and tried to stay just a few pages ahead of the students in all of my classes. It seemed as though the students purposefully tried to expose my ignorance by asking impossible questions during every class. They made sport of all my shortcomings, picking on my Southern accent, correcting my English, and being hyperconscious of the smallest mistakes, such as misspelling a word on the board.

Every raised hand seemed like a challenge to my authority as a teacher and if I had learned one thing from my previous year of teaching, it was that a teacher without authority and student respect could never teach, no matter how prepared she was for a lesson. Filled with the memories of just how out of control a class could become, I was determined to gain control of the situation. In doing so, I completely overreacted. Even at the end of my third year of teaching at Urban Magnet, students still recall those first few months when I spoke like a drill sergeant, making some of them actually cry from fear. I laid down the law, trying to gain control of issues that had been of particular importance in my old school, not necessarily realizing the same issues might not be as important here. I had somehow become so immersed in the culture of my last school, that I did not consider that the culture of this school could be different and would require different tools if I were to be an effective teacher.

What Do You Mean You Don't Know?

Well into the third chapter of our chemistry text and in the middle of a class discussion about the nuances of wave theory as related to quantum mechanics, I realized I was completely unprepared for and unable to answer the types of questions my students were asking. Although difficult for me to say at the time, I finally threw up my hands and declared that I "have no idea how to answer your questions, but let me write them down and I will try to find out!" With a look of utter disbelief, one of the students exclaimed, "But you're the teacher, what do you mean you don't know?" His arrogance to ask such a question and his naïveté to expect someone could know everything actually helped put the situation in perspective for me. For the first time, I realized that I was arrogant to assume I could know everything and be prepared for all their questions. It was at this point that I finally let down my guard at Urban Magnet and actually took the first steps toward building a learning community. I realized that I was not being sincere. I had allowed my students and me to accept the fallacy that I knew everything and all they needed to do was listen to me in order to learn. I explained to the students that they were as capable as I was and that they could make as much sense from the book as I could. I told them I worked the problems each night just like they did, many times making the same mistakes as I went along. I admitted that many of them were undoubtedly more talented mathematically than I was and for that matter, even more intelligent. I continued, telling them that what was important was that I was trying my best and as soon as they realized that teachers were people who sometimes made mistakes, the sooner they would be able to benefit from my teaching.

From that point on, I began to ask if any student could better explain a concept on which I was unclear and if I was unable to solve problems, I asked students to show me and others how they had solved them. This was not an easy thing for me to do. I had to learn to put aside my ego and pride, and I was rewarded by richer class discussions and feelings of camaraderie that developed in the class. Certainly some students still made pointed remarks about my shortcomings as a chemistry teacher, but I rarely had to come to my own defense as other students were quick to point out that I was doing my best and that was what mattered. Indeed, I have never experienced the same sense of kinship with later chemistry classes as I did with my initial class.

A Natural Progression

What happened next seemed like the only appropriate step to be taken. If I were going to teach chemistry well, I needed to learn more chemistry myself. The opportunity presented itself to me with an invitation to join the new Master's of Chemistry and Education (MCE) program at the University of Pennsylvania. I applied near the end of my first year of teaching at Urban Magnet with the knowledge and support of my chemistry class. My students were impressed by my desire to continue my education so I could better support their learning. Before the required taping of my class for the application process, students held an impromptu meeting, making certain the tape would reflect our best class effort for the admission committee. Upon acceptance to the program, my students congratulated me and in the years to follow, they continued to ask about my progress in the course and to express their support for my continuing education. My determination to return to school in order to better teach chemistry allowed me to gain social capital in a different way at Urban Magnet than at Dewey Middle. I began to realize there are different ways of gaining social capital with students and that it is necessary to do so in order to build relationships with students so that I could teach and they could learn.

Becoming a Real Teacher

A New Approach

My experiences in the MCE program radically changed my perceptions of what it means to "know and do" chemistry and what it means to teach and learn chemistry. My involvement with the program allowed me to question traditional views of learning and science associated with previous experiences. One professor in particular made a profound impact on my ideas about the teaching and learning of chemistry. Professor Johnson exposed me to models of teaching and learning that were different than I had experienced before. His use of an inquiry model introduced me to teaching and learning that emphasizes prior student knowledge and construction of student understanding and community learning. My own gains in knowledge and experience in chemistry and my interactions with Professor Johnson have required me to take on a new role as a learner. This became significant in two ways in my classroom; I began to teach chemistry differently—relying heavily on group learning and student construction of knowledge and I had a better understanding of what it means to be a student in a chemistry class. I am more aware of students' fears and concerns surrounding the learning of chemistry and I am able to discuss my personal experiences of learning chemistry with them—again building a sense of community in learning. I am better able to understand their frustration with chemistry concepts and can provide examples from my own learning to help them find new ways to tackle tough problems. Professor Johnson was also fond of incorporating a historical perspective, presenting the lives of scientists and the importance of new inventions in chemistry, making chemistry relate more to the real world and making me feel more in touch with the world of chemistry. My experiences with him as a teacher and as a learner in his class provided me with more knowledge of teaching and learning which I am now able to enact as I teach. Students often comment that they feel more in touch with chemistry when they learn concepts within the historical context of the discoveries we are exploring.

Having completed the MCE program, I feel more confident in my classroom and with the subject matter in general. I am comfortable with the idea that I do not know everything and I am confident in my ability to find out what I need to know in order to best support the learning of my students—including having my students explore a problem and suggest their own solutions. I am more certain of my ability to lead my class toward our goal of learning chemistry, all the while realizing that I am continuing to learn every day. My confidence in the subject matter allows me to do many things I was afraid to try before, like stray from the confines of the book or freely and openly explore student questions without the fear of not knowing the right answers. Being able to freely admit that I, as a teacher, am not privy to all knowledge allows me the freedom to seek additional resources both within and outside of my classroom. Professor Johnson, who is a senior scholar at an Ivy League school, has demonstrated his extensive knowledge of chemistry and his ability to say he is not sure, and will look it up. Both practices have inspired me to be a better teacher and a more willing student in his class and I have been able to transfer this experience into my classroom for the benefit of my own students. Professor Johnson allowed me to experience a teaching and learning style that relies heavily on community effort and I have found it to be an empowering experience for myself as a learner and for student learners in the classroom. Just as students in Professor Johnson's class feel empowered to ask questions, so do my own students because they know I value their knowledge and I treat them as my intellectual equals.

Looking Back

This autobiographical account of my journey of *becoming* a teacher has provided an opportunity to reflect upon my teaching from a new perspective and to grow from the experience. What I have learned goes beyond individual experience and has far-reaching implications for what it takes to be an effective science teacher in two profoundly different urban schools. Several themes have emerged from this historical analysis of my life as teacher and student that have been influential in shaping the teacher I am today.

Teacher Knowledge

As a child, I learned that teacher knowledge was the most important resource for learning in the classroom. My family and I regarded teachers as infallible. They were expert in all facets of life, holding the key to the world of success like a shiny carrot for me to follow. When I became a teacher, I had the same expectations for myself and I was overwhelmed by my seeming inability to perform as this perfect person I felt a teacher had to be. However, my experiences in both of my teaching environments and as a student in Professor Johnson's class convinced me that no one person could possibly know everything. This allowed me to identify my shortcomings as a teacher as well as admit them to my students. Only then was I able to grow and become a better teacher. It is unfair and unrealistic for teachers to be portrayed as the holders of all knowledge, effectively robbing students of their contributions to their own learning and to that of their peers.

Power in the Classroom

I have learned that a real teacher promotes student learning in a variety of ways, and not all of them are safe for a teacher who is unable to admit her/his inability to know everything. Moreover, while relinquishing control of teaching allows students to actively participate in their own learning of science, student learning is most enhanced when they are encouraged to use their cultural capital as a resource for learning. This can be accomplished by promoting student knowledge as important and valued in the classroom. Group-focused learning and students acting as teachers are two ways in which students were empowered in my classrooms at Urban Magnet. I encouraged students to teach chemistry concepts to one another in small groups as well as to the class as an individual or by utilizing group efforts. I found that allowing students opportunities to teach demonstrates to them that they also hold knowledge and can share information that is constructive and beneficial with their peers *and* their teacher. In this manner, my students began to see themselves as coteachers and to view me as a colearner in the classroom, thus elevating students to a position of power and de-emphasizing the teacher's role as the bearer of all knowledge.

Valuing the Culture of Students/Diminishing the Role of Other

Irrespective of the situation, my experiences have revealed that building social capital is critical to being an effective teacher. Recognizing student cultural capital as worthwhile and identifying what students know and can do as legitimate demonstrates a respect for students that allows their identities as learners to grow. By building social capital with students, I was able to form cohesive learning environments that acted to acknowledge and value their contributions to the learning environment. While the route toward building social capital is to *earn* the respect of others in a community, I discovered that earning respect in a community is clearly dependent on the type of community. At both Dewey Middle and Urban Magnet I had to find out what was valued and practiced in a community in order to gain the respect of my students. I learned that earning, not demanding or expecting, respect is a critical ingredient to becoming an effective teacher. Earning respect allowed me to be accepted by students as their teacher so that learning could begin. Furthermore, earning respect required that I enact roles appropriate for *those* students in *that* classroom since a classroom culture is largely dependent upon what students bring from their lifeworlds, not only on what the teacher brings to the classroom. By respecting student cultures and building a curriculum that focuses on valuing student knowledge, the feeling of "Other" by marginalized students is reduced within a community that emphasizes learning for all while relying on the resources contributed by all.

The Power of Autobiography

When first given the task to write autobiographically about my teaching experiences, I had no idea how influential the experience would be on my teaching practice and my view of myself as a teacher and a learner. The process now reminds me of watching my grandmother while quilting. After observing her gather odd squares of material and thread them together, I would sit and trace the outline of intricate patterns she had created with leftover materials from favorite shirts and clothes from family members. These "memory quilts," small bits taken from our lives, were then all threaded together to make one coherent piece. Like the memory quilts, pieces of my history alone do not seem very meaningful, yet tied all together, they present a powerful story of the person that I have become.

This autobiographical writing of my journey to become a teacher has been empowering because it has provided me with a lens through which to view my past educational experiences and their relative impact upon my experiences as a teacher and a learner. This journey has provided me with an opportunity to share the role of education in interrupting social reproductive cycles in my life and how I realized my own agency in the process. Much of what is written about teaching is historically removed from teachers and is often presented in a narrow context. In contrast, autobiographical writing is socially and culturally grounded in life history, a fact that enabled me to make connections between my past and the present as a means to continually improve my teaching practices. Autobiography can serve as an effective tool for examining and improving teaching practice because it challenges an individual to consider not only what students bring to the classroom, but what you, as the teacher, bring to the classroom. It is by understanding all facets of oneself as a teacher, a student, and a person that allows the successful building of learning communities that benefit both the students and the teacher. It is in knowing *how* you have become a teacher that you can begin to understand *why*.

Editors' Perspectives

The practices of a teacher are historically constituted. Accordingly, autobiography is a way to situate teaching in the experiences teachers have had with learning and teaching and to locate their social and cultural lives on a trajectory that depicts their positions in social space as a function of time. However, writing an autobiography is quite a challenge because it is not clear what to include and what to leave out. In this case emergent themes from Martin's classroom ethnography suggested some elements to include in her autobiography and elements of her autobiography suggested what to look for in the research.

The autobiography depicts Martin as growing up in conditions of poverty and struggling not only to learn but also to overcome her social and cultural otherness and gain access to resources to support her continuing education. Although Martin was challenged in her initial year of teaching at Dewey Middle school, it is apparent that her capital, derived from experiences with poverty and making do, equipped her well to adapt to the capital of her urban students. As she had done throughout her life she never gave up, thereby earning symbolic capital with some students and creating opportunities to build social networks and gain even more symbolic capital, in the form of students' respect. Martin created solidarity with many of her middle school students and was reluctant to leave them to go and teach at Urban Magnet. To this day she has continued to support her middle school students and assist them to improve their lives through education. She's *got the backs* of her students and supports them in many facets of their lives. In one notable case she coached a former Dewey Middle school student to succeed on her SAT test, employed her as a student researcher in the past summer, assisted her to gain admission to a prestigious college, and drove her several hundred miles to start her freshman year.

Chapter 14

Beyond Either-Or: Reconsidering Resources in Terms of Structures

Sarah-Kate LaVan

This companion chapter describes research carried out in Sonya Martin's tenth grade chemistry classroom at Urban Magnet and emerges from the collaborative research that she and I participated in. Using the lens of cultural sociology (Tobin, chapter 2) I foreground the dialectical and recursive nature of structure and examine how the schema and practices from participants' lifeworlds serve as resources for the teaching and learning of chemistry in Sonya's classroom. Specifically, I address the following questions: What resources are available to support teaching and learning in Sonya's chemistry classroom? Who appropriates these resources? In what ways are these resources important for students from diverse backgrounds to meet individual and collective goals? In considering these questions I reject the notion that you either have resources or you don't, and move beyond this false dichotomy. This is possible if we recognize that each of us always has resources, that is, schema and practices that we use each day to structure the fields in which we act and through which we create and appropriate new resources.

In recent years, research has cited failing test scores, heightened dropout rates, and a dwindling interest in science to illustrate the persistence and even the expansion of inequity in science achievement along social, ethnic, and racial lines (Seiler, 2002). Tobin, Seiler, and Walls (1999) have noted that African

American students from low-income and urban backgrounds often contend with out-of-date textbooks, limited access to laboratory equipment, and teachers lacking proper qualifications. Although there have been many reasons proposed for this urban achievement gap (e.g., declining budgets, teacher shortages, and lack of resources) the general calls for improvement in science education have been articulated through the position that all students will achieve given additional resources, tools, and support structures (National Research Council, 1996). As a result, policy makers and curriculum designers, intentionally and unintentionally, have categorized low socioeconomic and urban schools as not providing students with resources and opportunities needed to succeed and have regarded material resources as the "cure" for the many maladies affecting science education.

Many teachers and researchers, including Sonya, have begun to understand that physical resources are only part of the structure of a field. Sonya often commented that, during her first years of teaching chemistry, she experienced difficulties in planning and implementing laboratory activities that were not only due to her lack of material resources, but also attributable to the power structure set up within the school as well as students' beliefs about her knowledge and what success in chemistry required. Sonya additionally explained that her own understandings associated with using laboratory activities and material resources were so limited that she could neither use nor adapt the resources that were available in ways that would allow the students to develop practices and schema associated with science. As a result, Sonya rarely tried new activities that required her to put her identity as a teacher on the line. In this way, the use of material resources in Sonya's classroom was reflexively linked to the culture that she held in her toolkit (i.e., schema and practices) (Swidler, 1986) as well as her identity in the chemistry classroom. Thus, just as material resources were inextricably linked to Sonya's access to human and symbolic resources the reverse was also true.

The reflexive and dialectical nature of schema, practices, and human and material resources make it impossible to examine the use and appropriation of resources independently. Therefore, centering this paper around vignettes allows me to consider not only the various types of resources that were available to Sonya's students, but also the interplay among those resources, how they were appropriated by individuals and the community, and the structures that afforded participants the use of those resources to meet their own goals of learning science.

In examining the resources I identified in Sonya's chemistry classroom I argue that having resources is not an either-or situation. As Sonya and her students clearly illustrate, if one reconsiders resources available to both the teacher and students in terms of the various school and classroom structures it is possible to identify and use resources that were previously thought to be unavailable or unusable. Additionally, viewing resources in this manner allowed Sonya to examine the resources that participants have traditionally viewed as valuable for the teaching and learning of science, and gain insights into how participants appropriate resources to meet individual and collective goals.

Capturing Activity in the Classroom

Drawing from a variety of ethnographic data, including Sonya's autobiographical account, interviews, and videotape of classroom activities, this chapter highlights only some of the findings that arose from eight months of research I undertook in Sonya's classroom as a participant observer and coteacher. Aligned with the other chapters in this volume, critical ethnography and collaborative research (see Seiler and Elmesky, chapter 1) provided the pathway to identify, document, and analyze teaching and learning practices from a variety of perspectives as well as restructure the traditional researcher/researched roles. A hermeneutic-dialectical process was utilized to select student researchers with differing socio-economic backgrounds and schooling experiences to participate in the research activities, including interviews and cogenerative dialogues. For example, the first student researcher chosen was a high-achieving female of Cambodian-Chinese descent who possessed little social capital in this classroom, and the second student researcher was an African American female with high social capital and relatively low science achievement. Thus, the use of a dialectical process allowed us to obtain feedback from various types and levels of learners on how their experiences with traditional science classrooms afforded the appropriation of specific types of resources by a diversity of learners.

Student researchers gathered data about resources and teaching strategies through traditional ethnographic means including recording of classroom observations in journal entries, interviewing class members, and analyzing videotapes. As the study progressed, many other students from the class showed interest in becoming part of the research. By the end of the study, every student in the class was involved in some manner, such as interviewing other students, interacting in cogenerative dialogues, or making a video ethnography. Furthermore, Sonya and I gathered data by interviewing students, occasionally providing questionnaires to obtain anonymous self-reported responses to questions, watching videotapes for salient events or issues, and maintaining field notes and journals.

Standard classes and laboratory periods were videotaped two to three times a week. Classes were usually recorded, from varying positions within the classroom, depending on the particular activities and questions that needed to be answered that day. Due to the participant nature of this research, it was often necessary to allow students to record activity or to place the camera in one location within the classroom for the entire class period.

Cogenerative dialogues, conversations between stakeholders about shared experiences (Roth and Tobin, 2002), were central activities (and fields) that had the potential to transform Sonya's science classroom as participants identified a greater variety of resources and reached agreement on plans to equitably access and appropriate them. The varying schedules of participants allowed us to conduct one to two cogenerative dialogues each week. Cogenerative dialogues focused on issues arising from the research and were facilitated by excerpts selected from videotapes of classroom activity. We viewed the tapes at various speeds, engaging in micro- and meso-level analyses in which we identified patterns of coherence and contradictions surrounding the use of resources in the classroom. During cogenerative dialogues participants identified and reviewed practices and associated roles of the teacher and students, especially those that were unintended and unconscious, discussed power relationships and roles of all stakeholders, and examined factors that mediated the use of resources. Participants created new social networks and built symbolic capital through their interactions with different members in the community (i.e., students, teachers, and researchers). The power of cogenerative dialogue lay in the ability for all participants to be represented in examining classroom activity and reaching collective decisions on assuming responsibility for the quality of classroom life. No longer was it regarded only as the teacher's responsibility to produce and maintain productive learning environments. Instead, all participants assumed co-responsibility for collective solutions and regarded cogenerative dialogues as a field in which problems could be identified and potential resolutions could be cogenerated. Furthermore, culture produced in cogenerative dialogues (i.e., schema and practices) was available for future use in other fields, including the science classroom.

School Structures as Resources

Cultural sociology sensitizes us to the possibility that the physical and ideological structures established by the school as well as those afforded by Sonya's own beliefs shaped the availability and use of material resources for her chemistry classroom as well as student interactions. Therefore, in exploring the resources that made Sonya's classroom a successful community, it was important to situate the classroom and Sonya's interactions with the students historically and socially within the school community.

Urban Magnet is not a typical inner-city school since students and teachers are selected to become part of the school community based on performance and reputation. As a result Urban Magnet places a great emphasis on and commitment to maintaining its exceptionally high performances on national and state standardized tests, numerous awards, and the placement of almost its entire graduating student population into four-year colleges and universities. As a result the school community generally accepted and promoted a culture of high achievement, reified through selection and retention practices, which many students from diverse backgrounds could not meet.

Although in her autobiography Sonya emphasized an abundance of material resources at Urban Magnet, contradictions were evident in the distribution of resources to particular teachers and students. Urban Magnet, which houses students from grades five through twelve, was originally the site of the oldest girls' public high school in the city. Since the girls' high school was built to accommodate fewer students than the school now enrolled, the physical building could not adequately house the population of middle and high school students and meet the specialized needs for science rooms. Accordingly, many science classes are held in classrooms that were not designed for and do not support activities that are traditionally conducted in science courses. In particular, the classroom where Sonya and her tenth grade chemistry class of twenty-nine students met contained only one lab bench (with no fume hood and only six laboratory stations), one sink with running water, and equipment and materials for about one-third of the students to participate in a laboratory activity.

In cogenerative dialogues (see LaVan and Beers, chapter 8), students and Sonya identified an inequitable distribution of resources to support science teaching and learning. Equity issues arose because of the assignment of teachers and students to specialized or nonspecialized science classrooms and a tendency for some veteran teachers to hoard materials and equipment and not share specialized equipment with teachers they regarded as unqualified. Hence, Sonya's agency to teach science as she wished was truncated by the practices of administrators and more experienced colleagues. As was the case with Sonya, teachers having the least qualifications and experience received the fewest materials and equipment. Often when Sonya made requests to borrow materials or obtain assistance from more experienced colleagues, she was denied access and was insulted by degrading statements about her limited teaching qualifications and lack of understanding of chemistry. In one conversation, Sonya explained,

> All of the materials and resources were locked away in Ms. Jordan's classroom. She was the only one who had a key to her room and would lock it at night so I couldn't get access to any of her things. Even the cleaning staff didn't have a key. When I wanted to use something I had to go to Ms. Jordan and ask politely if I could borrow it. If she thought that I wouldn't be able to handle using the equipment or materials, she wouldn't allow me to borrow it.

How a Plunger Works

Even though physical and human structures (e.g., physical layout of the classroom, insufficient material resources, and Sonya's chemistry/math understandings) constrained the teaching and learning of chemistry in Sonya's classroom, Sonya's ideologies surrounding the students and the selection of tasks/problems as well as the role I held as coteacher and coplanner and my background in the university chemistry program allowed Sonya to alter the classroom structures so that all students thought, wrote, manipulated, and did science. Sonya accomplished this by restructuring so that resources she could access could support individual and collective interest, comfort, and agency.

Sonya was exposed to the use of collaborative peer groups in science teaching through the Master's in Chemistry Education program in which she was enrolled. Although not the focus of chapter 13, the uses of collaborative peer groups and inquiry questions were important pedagogical strategies that Sonya added to her repertoire, and peer groups soon became important in the classroom for both Sonya and the students. In comments received at the end of the year some students attributed much of their success in chemistry to the comfort they felt being able to co-construct knowledge and understandings with members of their peer groups. According to Sonya, the small, intimate nature of peer groups provided students with greater access to human resources, discourse, and active learning.

When the use of collaborative peer groups began, Sonya and I both anticipated lots of side conversations and off-task talk while the students were working on their assignments in groups. However, what we commonly found was that most groups were focused on the task at hand and students were actively explaining concepts to each other, making assertions, and using science language to describe their ideas. The following vignette illustrates the complex and varied interactions that took place within one peer group on an occasion when the students had been asked to apply their understandings of Boyle's Law to describe the principles behind how a plunger works to clear a clogged toilet. It is important to note that while these interactions are representative of those that occurred in this classroom, the practices, schema, and physical resources participants drew on depended to a significant degree on the culture available in the participants' cultural toolkits (Swidler, 1986).

In the videotape of this interaction the students were gathered around one piece of paper, on which Anna had drawn a diagram of the toilet bowl and the clogged pipe. Three of the students (Abby, Anna, and Joe) were seated at the table close to the piece of paper, while the fourth (Debra) was standing behind Joe and Anna, leaning and looking over their shoulders.

Speaker	Language	Gestures
Debra	What are you trying to find out? Why it works?	*Joe slides the paper closer to him and points to the diagram. Joe then shares the diagram with Anna.*
Anna	=We're trying to find out exactly *why does that *come back up. (1.0) Like we know what *you said.	**With her pen points to the clog on the diagram* *slides the pen up the pipes to show the clog coming out.* *Turns head slightly to left toward Debra. Debra's hand comes to her chin and scratches and looks at the paper.*
Joe	What, what?	*Joe, looks at the paper. As he talks Anna is the only one who looks at him. Other two continue to look at the paper.*
	Long silence (3.7) from the group members	*Joe repositions himself and leans his upper torso on the table so that his head is resting on his arm. Joe is now directly in front on the paper with pen in hand.*
Debra	Well (0.4) maybe it's when you *push down	*Standing slightly behind the other students, *makes motion of pushing a plunger down with her left hand. Group members are looking at the paper while Debra gestures.*
	[and the pressure um	*Hands still in plunging position.*
Anna	[The thing is that gets me is that when you, when you	*Raises her head slightly to the left toward Debra as she begins to speak. Points to the diagram*
	[*push down and you	*Head still raised. *Makes motions of plunger with both hands.*
Joe	[When you pull	*Looks at Anna. Lifts his body up from the table.*
Debra	=When you apply more	*Makes the motion of a plunger with left*

	pressure	hand. *(Same motion as before)* Anna and Joe watch.
Anna	=When you pull when you pull back up,	As Debra finishes plunging, Anna looks up to her and takes both hands and *makes two small gestures showing the pulling back up of the plunger.*
	the same pressure that <u>was</u> there *before is <u>there</u> again.	Pulls paper closer to her. *Points to the paper with left hand four times while looking at Joe. Then pulls hand away. Debra watches closely.*
Debra	=Yeah↓	Moves her hand to her face, in the pose of the thinking man statue.
Joe	Right↓	
Anna	So then why↑	

Collaborative Peer Group Structures as Resources

It is evident from this vignette that peer interaction and characteristics of the group itself played important scaffolding roles for the learning of science. Sonya and I often discussed in cogenerative dialogues the nature of the assigned problems/tasks, their challenge to students and requirement that individuals apply their knowledge and skills to form larger conceptual understandings. The structures set up by the problems themselves and use of peer groups encouraged sustained discourse, distributed expertise, and collective leadership within the group and the class, and thus led to the valuing of human resources. The uncertainty that each individual exhibited as well as the co-construction of ideas and continual questioning of each other and themselves illuminated incomplete understanding, and necessitated reliance on one another for learning. Anna's statements about her partial understanding and her assertions about the pressure before and after a person operates a plunger were not only supported by her group members, but these declarations also provided Anna with validation that understanding only a piece of the puzzle was acceptable. Additionally, these comments afforded Anna the structures necessary to take the group's understanding to the next level by asking the subsequent question of why the pressure is the same. In this way, Anna's question served as a resource for her own understanding and for the group's as well.

Also significant is the distribution of power and mutual focus within this group, where all participants seemed to have a shared responsibility for making meaning and contributing to the process. In the transcript, this was illustrated through the students building on one another's ideas and sustained focus or train of thought that was generated. For example, when Debra asked, "Well, maybe it's when you push down," Anna jumped right in and stated that she was stumped by what happens when someone pushes the plunger down. Debra quickly corrected her own language, and that of Anna, stating that the phenomenon occurs when "you apply more pressure." Anna, without hesitation, seamlessly developed the concept further, "when you pull back up, the same pressure that was there before is there again." This exemplifies Hogan's (1999) notion of co-constructive talk or "seamless contributions" as a necessary resource for gaining science fluency and dispositions, such as asking questions, making assertions, and building arguments for answers.

Student Generated Resources and Understandings

In addition to human resources, the students utilized various other resources to help solve the problems assigned to them in their peer groups, including but not limited to textbooks, physical objects such as a plunger and toilet, models, class notes, and understandings of chemical concepts associated with gases. However, when the activity began, the students in this particular group were at a loss as to how to begin their discussion. Questioning Sonya about where to begin their inquiry, she suggested that the students explore the various resources available to help them get started. Feeling that a diagram of the toilet and plumbing would help her begin to understand what was going on, Anna drew a diagram of the toilet and plumbing on a piece of notebook paper and placed it before the group of students.

Anna's diagram of the toilet and associated plumbing served as an important tool in the development of the schema and practices of the group and focused the interactions that took place. No longer than two seconds after Anna placed the diagram in the middle of the table Debra asked for clarification about the group's problem and objectives. Immediately following Debra's statement, Anna explained that they, as a group, needed to figure out "why that (pointing to the clog) comes back up." Thus, Anna's sketch became an important visual aid, served as a referent for the students' attention, and functioned as a symbol from which to build shared language and understandings about the problem. The students referred to the diagram and the image of the plunger, throughout their discussion, and in the reporting phase of their inquiry, they and others used the diagram as a symbolic resource with which to incorporate culture they brought to the classroom, discuss the concepts, and negotiate understanding.

Though Sonya intended for students to work as a collective with shared responsibility for learning, students ultimately were afforded individual agency to decide which science knowledge they would draw from as well as how they would proceed with activities and learning. As a result, students often took different directions once peer group learning was underway. The amount of collaboration and participation within the group was variable and dependent on the dispositions of the individuals, the tools and culture the students brought to the group, and other materials that they appropriated either individually or collectively. This was clearly illustrated when Joe, who was having difficulty "seeing" the phenomenon, suggested the group obtain an actual plunger to examine the pressure relationships. When the group denied his request, Joe went against their conclusion that a plunger was not necessary and obtained one from the custodial staff. When Joe returned to the table, his group members questioned him about the plunger and asked to examine the space inside the bulb so that they could confirm their theory about the amount of pressure present before and after plunging. Unlike other occasions when Joe interacted with the group, his control of this artifact (or tool) allowed him to become a central resource, to whom various group members addressed questions and comments. This altered structure provided Joe with greater opportunities to explain and manipulate ideas as well as to draw on his experiences of using a plunger.

Joe's acquisition of the plunger served as a resource for the other members of his group during the small group activity as well as for the remainder of the class during the reporting stage of the activity. Science concepts and resources are often unfamiliar and removed from the everyday experiences of students therefore the plunger also provided a few of the group members (specifically Joe and Anna) with a resource that gave them comfort and familiarity. Setting up structures in which students felt comfortable and could see the relevance of learning was important to Sonya in creating an environment in which students could begin to take risks, share ideas with one another, and have meaningful science conversations. Thus, having students access and appropriate resources with which they were familiar was useful in obtaining buy-in and having students begin to connect with science.

Speaker	Language	Gestures
Joe	And when you pull, when you pull, [and	*Points to diagram with left hand. Anna and Abby look directly where Joe is pointing.*
Debra	[So, so when you push *down,	*Leans over Joe to be able to point to the diagram. Joe moves his hand over and leans down just before Debra places her hand on the paper. Debra *makes plunger motion and then pulls her hand and body away and pounds fist to open hand three times.*
	you're increasing the pressure, which pushes <u>that</u>	*Debra leans forward and *points to the diagram. Debra leans out again and when she*

	out	*does this Anna and Joe who are sitting next to each other look at each other and smile.*
Anna	=No! It comes* back up again.	*Looks at and *points to the diagram. Looks at Joe.*
Joe	Cause, *cause when you [pull	*Joe raises his head and looks at Debra. *Puts left hand on diagram then takes pen (in right hand)*
Anna	[Have you ever plunged a toilet? (*Laughing*)	*Before speaking, turns head around to Debra who is standing behind her to the left. Anna asks the question, smiles and then breaks into a laugh. Debra smiles. While smiling, turns her head back around and looks at the paper.*
Joe	[Because when you pull a plunger out, water comes I mean air comes back up and everything comes back up and then it all goes back down.	*Moves right to left on the diagram. Makes sweeping motion with left hand as he takes the pen away from the diagram. Leans head and body back on arms. Debra has hand to face in thinking pose.*
Anna	Yeah. And then it washes back down.	*Holding left index finger about six inches above paper points down at diagram. Then looks up to Debra.*
Joe	Right. So why does it all come back up?	*Leaning on table.*
Debra	Well, is there a vacuum created in *there?	*Leans across Joe and Anna and *quickly points twice to the diagram*
	When you *push it down does it become a vacuum? (2.0)	*Pulls hand away and *makes two quick plunging gestures. Anna drops pencil and Anna and Abby reach to pick it up. Debra looks quickly at them.*
	Because when you let go all of the *air would be sucked back up?	*Debra reaches across Joe and Anna and *points to diagram and brings hand back to face in thinking position.*
Abby	There's no sucking in science! (*Laughing*)	*Looks at Debra and smiles. Group members smile and laugh.*
Debra	I mean it would be pulled.	*Smiles.*

Language and Gestures as Resources

During this interaction sequence students used canonical and everyday language, gestures, and body orientations in the negotiation of meaning. It is apparent that any given action was a resource not only for the subsequent learning of the actor, but also for any member of the group. The use of symbolic metaphor, substituting "that" for the clog, the frequent reference to the diagram, and the common usage of canonical science concepts such as pressure and vacuum, show that the students held some shared understanding and language from which to approach the problem. However, it is also evident that they did not share identical cultural capital, and thus did not begin the problem from the same point of view or draw on the same symbolic resources. Whereas Debra began from the very foundation of the problem, noted by her question, "What are you trying to find out?" Anna skipped that first step, and proceeded directly to drawing on understandings the group had established the previous day. Furthermore, while Joe and Anna were able to draw on their real-world experiences with a plunger, they joked with Debra and suggested that her inability to do so prevented her from understanding the physical chemistry of the plunger.

Through an examination of Debra's actions we can begin to understand the significance of language, gestures, and body positions as resources that structured agency and learning; participation and understanding of the students within the group. Due to Debra's spatial orientation to the group (i.e., she was behind them), the three other students could not see her gestures and facial expressions. Therefore, in order for Debra to illustrate her understandings and communicate clearly with the others, either the students had to turn to face her (which happened in one situation) or Debra had to make herself "seen" by reaching over Joe and Anna and stretching her body horizontally toward the diagram. In this vignette, although Debra appropriated the symbolic gesture of a plunger, it did not aid in making her understanding clear to the others. Only when Debra chose to access the diagram by reaching across Joe and Anna and coming into the others' fields of vision was she able to use the diagram, gestures, and language to articulate her understanding to the rest of the group.

Cultural Capital and Multiple Resources

What it is interesting and significant is the manner in which the students reflexively appropriated symbolic, human, and material resources. The frequent use of multiple types of resources served not only as common referents from which to structure the activity but also allowed the students to express ideas, test out understandings, and work toward developing a larger understanding of the concepts. In Sonya's classroom, students were often encouraged to refer to multiple resources, including their cultural capital, to aid them in making connections and understanding the concepts. In videotapes Sonya was often heard repeating the phrases "Smart people use their resources. Even scientists use their resources. They can't know everything," to get across her beliefs that accessing and appropriating resources are important practices to enact in learning science.

Although Sonya mentioned in her autobiography that this strategy was developed out of her own necessity to use resources and as a means to push students to learn to think and not merely memorize details or concepts, she continued to employ it since she found that it allowed success for students who usually were unsuccessful in learning science. In the above transcript, we saw that certain structures created by Sonya and the peer groups stimulated resonances with students and allowed them to appropriate their cultural capital and symbolic resources in very different ways. Joe, being very familiar with the practice of plunging a toilet, was not only able to follow the chemical concepts and scientific argument put forth by his peers, but he was also able to use this knowledge to question others in order to gain greater understanding of chemistry. Additionally, Joe's explanation and question of "why does it all come back up?" prompted Debra to speculate "that there must be a vacuum created." This assertion became a resource that was instrumental in allowing the group to build consensus and understandings about the problem as well as report their findings back to the class. Thus, Joe, who was one of the lowest-performing students in the class at the beginning of the year, might have been excluded from the field of science had he not been afforded opportunities to build understandings from his own capital. Further, through interacting with peers in his group, he was not only able to achieve at a higher level, but also served as a resource for other students, helping them gain dispositions leading toward scientific fluency.

A series of interactions between Anna and Debra illustrate the complex interchanges of capital and the manner in which cultural production occurs in the learning of science. Anna felt that Debra was dominating the conversation and wanted to impress her with the salience to this problem of her own knowledge. Accordingly, she proceeded to taunt Debra, asking whether she had ever plunged a toilet; visibly rolling her eyes as she looked at Joe and moved her body in a deprecating manner. Her actions catalyzed laughter among the members of the group and allowed Anna to validate her everyday knowledge and gain social and symbolic capital by teasing Debra, who was regarded as knowing more science. These interactions were resources for others in the group. Before Anna made the comment, Debra was unequally appropriating turns at talk and her uses of the diagram. However, following Anna's comment Debra remained silent for some time. Anna's taunts made Debra laugh and take a step back, figuratively and literally. When she spoke again, Debra had formulated a tentative response to the problem. Anna's comments also served as an illustration to Abby that it was acceptable to correct Debra. Rarely when working in groups had students contradicted or refuted Debra's statements in the past, however Abby's exclamation that "There's no sucking in science" was a source of social and symbolic capital for Abby, since Debra accepted the remark as correct, appropriate, and humorous.

Becoming Fluent in Science

A large part of becoming fluent in science involves coming to know and understand the shared values and accepted methods of interaction (practices and dispositions) within the scientific community, such as argument, observation, and description (American Association for the Advancement of Science, 1993). The field-dependent nature of the culture of science (with its associated schema and practices) however, makes it almost impossible for a participant to move to a more central position within a field without interacting with others, oftentimes those who are already steeped in the culture (Lave and Wenger, 1991). Thus, the accessing and appropriating of human, symbolic, and material resources become crucial for novices to develop into fluent science participants.

All too frequently students in urban schools are placed in science classrooms with teachers who are unfamiliar with the culture of science or too afraid of science or the students to provide them with opportunities to do science. As Sonya articulated in her autobiography and in interviews, she found herself in a similar position, where limited chemical and mathematical understandings, difficult school structures, and ineffective pedagogical strategies prevented her students from learning chemistry. However, through fostering structures that promoted peer interactions, inquiry processes, distributed expertise, social networks, and valuing of cultural capital, Sonya and her students were able to recognize, access, and appropriate a greater variety of resources for learning. In turn this provided students with opportunities and structures that were previously not available.

The structures that constitute Sonya's chemistry classroom are socially and historically constituted, and shaped and reshaped by the physical resources, schema (e.g., beliefs about science, knowledge, teaching, and what students are capable of achieving), and practices (e.g., rule structures, gestures, and language) brought to the field by all stakeholders. The structures are mediated by a variety of factors including Sonya's beliefs about students interacting within peer groups, the capital that participants bring to the classroom, the presence or absence of physical resources, and the community's beliefs about Sonya's knowledge of chemistry. Additionally, this sociocultural lens suggests that since the negotiation and appropriation of resources are mediated by multiple factors, they are not limited to classroom interactions during periods of teaching and learning.

Thus, for Sonya, the networks she forms with colleagues at Urban Magnet and in her university chemistry program and the cogenerative dialogues (see LaVan and Beers, chapter 8) held with me are crucial for gaining additional science understandings, obtaining material resources, and as colleagues coplanning and bringing the knowledge of the university chemistry program into the students' science curriculum.

Similarly the structures set in place by administrators and other faculty members as well as the amount of chemistry knowledge that Sonya brought to the classroom shaped the types of resources that could be accessed and appropriated in her classroom by her students. Held in a dialectical relationship, the schema, practices, and resources associated with structures were dynamic and shaped students' practices, schema, and capacity to appropriate resources. As illustrated in the plunger vignette, certain structures resonated with tools that individuals held in their toolkits and allowed each to enact specific practices and use specific resources to meet their goals. For example, Joe used the structures set up by the plunger problem and peer groups to obtain a plunger as well as draw on his experiences with plunging toilets.

Educational implications from this viewpoint emphasize that in order for classrooms to become discourse communities, where transformative teaching and learning occur, we need to value classroom practices that change the typical power structure that exists and challenge the ideology that maintains that the teachers' culture and understandings are the primary sources of knowledge. It might help for teachers to consider learning as both an individual and social process, as well as enculturation into the domains of science, in which students are expected to access and appropriate resources (human, symbolic, and material) to meet their goals. Therefore, the teacher not only "knows science," but also can make the cultural tools of science available to learners, by supporting their reconstruction of the ideas through the culture they bring to the classroom and through discourse about shared physical events. Over time, cultural practices and schema change, thus affording no participant the ability to predict which new resources will be of greatest use. Therefore, it becomes important for participants, especially from populations who are often placed into oppressive cultural positions (Delpit, 1993), to be taught to identify, access, and appropriate resources to support their learning. The role of the teacher is to enact teaching practices that help to structure the learning community in ways that allow students to access cultural tools and dispositions they bring as well as those of the community, and in addition, to provide opportunities for students to make sense of these ideas and dispositions for themselves. In this manner, all participants are encouraged to view science as a fluid process of challenging and testing ideas and to understand that individuals are capable of transforming science and their lives.

Editors' Perspectives

A central part of our theoretical framework is a both | and perspective. When dichotomies are postulated we bring what might be considered the extremes into a dialectical or recursive relationship. This approach guards against the privileging of deficit views and encourages researchers to look beyond patterns of coherence to identify contradictions and explore the transformative potential of resolving contradictions. Structure | agency and individual | collective are two powerful dialectical relationships that serve to organize much of the research in this book.

Chapter 15

My Cultural Awakening
in the Classroom

Linda Loman

As the daughter of a teacher, I should have been better prepared for the life of a teacher. I had seen my mom preparing lessons, grading papers, and, at times, I even visited her class. I admired my mom and often considered following her footsteps, but when I got to college I felt I had to prove myself by selecting a challenging course of study. Other students portrayed education classes as simplistic and easy. I wanted to be able to say my major and have other students identify me as someone who pursued difficult classes. I majored in physics and had minors in math and German and considered going on to pursue my Ph.D. in physics. Yet after spending two summers as an intern at the Pacific National Laboratory and the Lawrence Livermore National Laboratory, I realized that I had no desire to live a life surrounded by equipment and a PC with only occasional collaboration with other human beings. I decided to become a teacher.

My journey toward becoming a teacher was perhaps not the typical course but it has given me a unique perspective on the career I took for granted. Perhaps because I valued the idea of becoming a prestigious experimental physicist and avoided taking education classes, I was not easily convinced that teaching was what I was meant to do. Yet as I worked and taught, I came to the conclusion that teaching was not only extremely challenging but also very fulfilling.

This paper consists of four sections that address my growth as a teacher in four distinct environments. First, I discuss my experiences as a novice teacher in the Peace Corps. Second, I describe my experiences as a short-term substitute teacher in the New Mexico school district in which I grew up. Third, I focus on my internship at an urban high school in Philadelphia. Finally, I examine my teaching at a magnet school in Philadelphia and my practices as a teacher researcher in my own classroom. I explore my development as a science educator through three themes: changing personal perceptions of my role as a teacher, my hopes and expectations of how my students' parents would be involved in their children's education, and how the availability and use of resources in the classroom affect learning.

During my second year of teaching in Philadelphia I was invited to work as a teacher researcher with a team of educators, led by Ken Tobin, exploring urban science education. Part of this work has included looking at my own teaching through different theoretical lenses, which allow me to understand teaching and learning in new ways. In my autobiographical account of experiences as a science educator in four very different settings, activity theory (Engeström, 1999) provides a sociohistorical perspective to my search for understanding. In an activity-theoretical perspective, actors' abilities to reach their goals are mediated by various factors in the activity system. For example in a classroom, efforts by teacher and students to meet their goals might be mediated by differences between the students, the tools available to support student learning, the manner in which rules are implemented, the roles played by the community, and the division of labor within the classroom. Often, exchange and sharing occurs between the different community members as they attempt to achieve specific outcomes. At times, objectives are not reached because obstacles and contradictions within the activity system arise between the various participants and factors in the activity system.

As one example of a contradiction, people may fail to meet their objectives because they do not have the necessary tools to achieve them. For example, a student may not have the language skills to be able to understand the teacher and therefore performs poorly in their class. Another student may not have the money to buy the necessary supplies, or the classroom itself may be lacking in equipment. In these situations, the lack of material or human resources presents an obstacle for both the teacher and the students in reaching collective and individual goals. Other contradictions can emerge when the cultural resources that the teacher and students bring to the teaching and learning of science differ. It is in these obstacles and contradictions that the possibility of change lies; that is, through understanding the contradictions we uncover the capacity for doing things differently.

One way in which such change potential can be found is through understanding the impact of building social and cultural capital with students. I use capital to mean something which a person can use to exchange or buy something of value. For example as I work to gain students' attention in class, I must have something to give to them in return. This might take the form of a discussion about a local sports team or a joke about riding the subways. My use of social and cultural capital allows me to achieve my goals of not only teaching science but also making meaningful connections with my students. I have learned that building and sharing social and cultural capital can reduce contradictions within my classroom. This can occur when members of the activity system find that they do not have the resources to meet their objectives, and must gain some sort of capital to exchange with others who can help them achieve their goals. For example, a teacher who learns to incorporate the culture of her students in the classroom can use this to connect with them so that she can achieve her goal of teaching science more effectively and increasing student learning. Furthermore, if a teacher is perceived as caring or even just "cool" she can use that symbolic capital to influence students to achieve academically, despite reputations that define the students as "troublemakers" in other environments.

I regard culture as intersecting systems of rules, beliefs, associated practices, and values. As a teacher I enact culture, as do my students, in order to attain goals related not only to their learning of science but also to their navigating life outside the classroom. These goals include my goals, my goals for my students, and their goals for themselves both in and out of school. These goals reflect our cultural values and, at times, are at odds with each other. My research involves an examination of my practices and my process of identifying the places where students use culture in the primary activity of learning science. Much of what I learned was directly related to the various environments and cultures in which I worked. As I adapted and grew as a teacher, I learned that many of my goals were the same in a variety of cultural settings, but the way in which my goals were achieved depended on the cultural capital my students expected to exchange with me.

The Toughest Job You'll Ever Love

I joined the Peace Corps in 1997; fresh out of college with my physics degree in hand, I shipped out with twenty-four other education volunteers to Papua New Guinea (PNG). I expected the teaching part of my assignment to be easy, and so

I focused on the challenges of learning a new language and culture. I received three months of in-country cultural and technical training. During that time, I lived with a family and traveled into the larger village of Goroka daily where I took language lessons, visited schools, and interned as a student teacher. I was then sent to Manus Island where I taught at Papitalai Secondary School for two years as a physics, mathematics, physical science, and physical education teacher.

Papitalai is a Catholic boarding school consisting of grades nine to twelve. Most classes had forty-five students with an even number of males and females. It was uncommon in PNG for girls to attend school, but the Sisters who ran the school were committed to providing girls an education in this male-dominated society. Education in PNG is not free and students must continue to receive good grades and pass national exams if they expect to make it to the next level. These requirements produce a learning community of highly motivated students each of which is supported by a large family unit, in the distance. Teachers and students spend the entire day together and are part of a tight community. Teachers not only teach their subject, but also serve as house directors and have duties that include supervising work and social events. The rules at Papitalai are very strict and include punishments that typically involve manual labor.

My initial idea that I would simply walk into a classroom and disseminate information quickly changed. These naïve ideas were based on my assumptions, often reflected in the culture of mainstream America, that teaching was an easy job with short hours and a long summer vacation. The logistics of teaching classes with forty-five students of varying abilities were staggering to a new teacher. In addition, my underestimation of the significance of language and cultural differences often led to miscommunication and misunderstanding. For example, when I asked Bonney, a ninth grade boy, if he had done his homework, he simply raised his eyebrows. I took this to mean that he wanted me to repeat my question, so I did. Again he raised his eyebrows. I felt he was testing me and I began to feel irritated as I repeated my question. Finally I made the connection that on Manus, a simple gesture such as a raised eyebrow meant "yes," while to me it had originally meant "what?" To further complicate matters, students treated me with a feeling of mistrust bordering at times on contempt. This was surprising since during my training I was assured that people would be happy to have my help at their school. However there was another teacher from an Australian Catholic volunteer service who had arrived shortly before me and had deeply offended the school community. Ultimately the Sisters and the principal dismissed him. Knowing this, I spent a lot of time proving that not all foreigners acted in an inappropriate manner. Thus, my initial efforts were not to simply teach kinematics, but to prove myself to be respectful of their culture and able to comply with their social structures. For example, there were several physical cues I had to relearn in order to avoid offending individuals, such as wearing long skirts, avoiding eye contact with most men, and not stepping over items. Although some people might question why I gave up parts of my own culture, I felt that the social capital I gained by taking on certain practices common to the women in PNG allowed me greater freedom in experiencing other aspects of their culture. Furthermore, after I had gained this social capital I was able to speak with them about sensitive subjects, such as abuse, without being dismissed as someone who did not understand what a woman in their culture experienced. In a culture where a woman's role is often secondary to a man's role, many of my male students, who were only three or four years younger than me, chafed at any attempts I made at classroom control. This was not something I had anticipated, and I was often forced to focus on classroom management instead of designing labs and activities. Only after several months of culture shock and management crises, was I able to focus on the actual teaching of science.

In a letter to my mother, I commented that every person should have to serve a year as a teacher, since too many people look at teaching as a simple job. In my first months as a teacher I realized that working with forty-five people simultaneously, while taking into account their changing moods and diverse backgrounds, was a monstrous task. I wrote that people would be more sympathetic and have a new appreciation for the diversity of people if they spent time in a classroom. On an occasion when one of my students was absent, I inquired when he would return. My students solemnly reported that his aunt had placed a curse on him, and until he appeased her he would not return to school. This frustrated me because I did not believe in the power of a "curse" and wished he would return to his studies. However, a fellow teacher and native Papua New Guinean, Sister Mary, told me that, even as a nun, she believed in such magic and suggested that I respect it even if I did not understand it.

By following the examples of the women in the community, I learned many of the social conventions and, in this way, I came to be accepted by my students as someone who respected their culture while maintaining my own culture. I taught the students about some of the values and norms of my culture, and shared my cultural capital in the form of physics knowledge. In this type of exchange, I was displaying specific parts of my cultural capital by teaching them science in ways that allowed them to learn. In my culture, this capital could be exchanged for jobs or further education. However, in their culture, where educational opportunities and jobs were limited, this currency was viewed as less valuable. Thus, as a teacher in PNG, I often found myself asking what value physics held in their lives. I could see that the ability to understand their surrounding world on a different level was valuable, and that provoking new questions and insights could not be harmful. But in the event that their future held less equations and more fishing, was this cultural capital really useful to them? This question is similar to those asked in urban centers where students often find contradictions between what they learn in school and what they need to survive in their neighborhoods. Even now when my students ask, "Why are we learning this?," I know that part of the reason is to reach my objective of teaching science. However, I also often find myself considering their objectives (which are diverse) and trying to decide whether the skills and knowledge gained in my class can help them reach any of their goals.

In PNG I lived in a very isolated community where there was little privacy from other students and teachers. Outside of the classroom, we lived as a small village where students, teachers, administrators, workers, nuns, and families all worked together. I often felt overwhelmed by the idea that I saw my students more often than their parents, who lived in far away villages. There was no such event as a parent-teacher night. Many parents never came to the school and a student's school life was kept separate from his or her village life. In their villages, students went by different names, spoke different languages, and they lived in huts made of natural materials, without plumbing and electricity. Furthermore, the rules at home were different than school and often caused them to deal with a variety of contradictions. For instance, at school, students were given punishments of physical labor if they were heard speaking a language other than English; yet, in the village, speaking English would have separated them from their families who only spoke their native languages. To help reconcile such contradictions, I thought that the community would benefit from a greater degree of parent interaction with the school, but I knew this was unlikely considering the distance of the school from the students' homes. Even so, I realized that in trying to understand the dual nature of my students' identities, I would have benefited from contact with their parents, and I carried this idea with me as I continued teaching, looking forward to a time when I would have greater access to my students' families.

Owing to its remoteness, PNG was lacking in resources and tools and this had a major influence on how I taught. I learned to teach using situations with which the students were familiar and to connect familiar contexts to what they were learning so that I could access their cultural capital and use it as a resource to mediate their learning of physics. I contemplated writing a lab book composed entirely of physics experiments that could be done with a coconut and a stopwatch. When teaching Newton's third law, we went to the ocean, only yards from our classroom, and used two canoes connected by a rope. I asked the students what would happen when Beno pulled on his end of the rope. Various answers were given. When Beno started to pull, students noticed he not only pulled the other canoe toward him but his canoe was also pulled towards Luanah. When we moved to topics such as Coulomb's Law it was more difficult to make connections to familiar contexts. Many experiments required electricity, which was provided by a generator that only ran sporadically. When we discussed changes of state, common materials such as water were used. However, some students had never seen ice and had little comprehension of what frozen water would be like. I had to make a conscious effort not to alienate my students from science when the most common examples drew from cultural experiences that were different from theirs. I would revisit this challenge later when I went to teach in an urban high school where many of the students felt science was a mystical field that had few direct connections to their lives.

Que Pasa?

When I returned from PNG, I lived at home while I decided where I would go next. I had enjoyed teaching and decided I would continue on this path. This surprised some of my friends who felt a Ph.D. in physics would be more worthwhile. However, I had found the challenges of my experiences exciting and decided that the rewards of teaching outweighed the struggles I had encountered. While applying to graduate programs, I worked as a substitute teacher in the Las Cruces Public School district in New Mexico.

While working as a substitute teacher, my growth as a science educator was hampered. There was little chance to work on skills such as creating new lessons, writing new labs, and using techniques specific to teaching scientific knowledge. I worked in a different classroom every day and taught every subject at every level. Students knew I was only there temporarily and I had very little social capital with them or with the other adults in the school. I realized that part of my success in PNG was due to my ability to cross the cultural divide that had initially separated me from my students. This commitment on my part translated into a sense of respect and trust from my students, enabling me to build social capital, which I could use in the teaching of science. In contrast, when I was working as a substitute in the high school where I was once a student, I did not have the opportunity to build the social capital that would have enabled me to teach at a more meaningful level. I missed the rewards of teaching that came from building relationships with students and other teachers.

My role as a teacher became centered around behavior management tasks and following directions left by the classroom teacher. I had little impact on deciding what activities would be used and how information would be taught. As a result, I started to feel disconnected from the students in the classroom, which was a new experience for me. I often felt at a disadvantage not speaking Spanish in a predominantly bilingual environment. In PNG I had used my knowledge of the language as cultural capital that allowed me to find acceptance in a culture that was not my own. However, in what was my native environment, I was often at a disadvantage because I could not always understand what students were saying. I had to rely on other avenues of building relationships during my short tenure as a sub. I tried to work in certain schools repeatedly. Once I became a familiar face and students knew they might see me again down the road, I held more social capital, which I soon learned to exchange for better classroom management. This reinforced the lessons I learned in the Peace Corps in that the actual teaching of facts was only a small part of my role as a teacher.

It was a strange sensation to work in my old high school and have lunch in the teacher's lounge with people who had known me as their student. I was amazed by the physical changes in the school and the implementation of new resources. There were new computer labs and a new wing to hold the growing student population. Other changes also surprised me. It was the first school year after Columbine and security and safety were on the minds of the entire community. Students were no longer allowed to go to their lockers, the classroom doors were locked between periods, and the school police roamed the building. I had only been away for eight years and yet I hardly recognized the daily routine, which included block scheduling. It amazed me how fast change occurred in schools. Most of the veteran teachers felt that these phases came and went in cycles and that there were relatively few real innovations in education. In effect, although the school had made major improvements in the available tools and resources, many of the rules had been changed and many privileges had been lost.

Life in the Big City

As a returned Peace Corps volunteer I decided to apply for a position in the Returned Peace Corps Volunteer Fellows program at Temple University in Philadelphia. The idea behind the program was to bring teachers with experience working in overseas resource-lacking environments into urban districts. Furthermore, this would be my first exposure to university classes on education. I interned the first semester at Neighborhood High School where I took over the responsibility of one physics class. Once again I was working in a culture completely unknown to me. Neighborhood High is a massive urban high school with about two thousand five hundred students, 99 percent of whom are African American. Urban schools such as this one are often characterized by a lack of qualified teachers, disconnected curricula, an absence of adequate resources, a deteriorating budget, and oppressive schooling practices (Elmesky, 2001). According to one of the teachers, only a minority of incoming freshmen would graduate in four years. Daily attendance was sporadic, the physical setting of the school was stark, and academic achievement in the school was low. The students came from a variety of neighborhoods throughout the city ranging from the projects to row homes in the area. I had never lived in a large city before, and I felt like a stranger in America.

During the first weeks of school I observed several teachers. Some were young and optimistic; others were older and seemed to have become jaded. The level of learning was very low as the teachers were forced to focus on control and safety within the class. In my journal I wrote:

> I take back all the complaints of things running late in PNG. It's 8:35, first period started at 8:00, and Mr. K is barely getting started (others haven't made it in yet). (Journal entry, January 2000)

I was surprised by the seeming disinterest of both the teachers and students. I spent most of my time with Mr. Torkana, an immigrant from Sarajevo, who had been teaching at Neighborhood High for the last ten years.

> According to Mr. B (the head of the science department) there are a few cultural and language obstacles in this class. The class has not come in and settled down, they are almost defiant. I need to really brainstorm about what it will be like to take over a class that has been used to running wild. (Journal entry, January 2000)

It turns out that these cultural and language obstacles were to my advantage. When I began teaching the class, most of the students were eager to learn and hear from someone they claimed could speak English. After my first few weeks I was feeling pretty optimistic with only minimal problems from a few girls. Then I gave my first test.

> OK, they didn't do well, only one person passed. Mrs. Barnes said she wasn't surprised because they never study. Today I'll be asking them to work harder and not give up. (Journal entry, February 2000)

With only one class to focus on, besides my coursework at Temple University, I was able to spend a lot of time on lesson plans and only a small amount of time marking papers. I saw my role as someone who could show these students that physics was not impossible for them to learn. I found that my experience in the Peace Corps often came in handy. One day I was explaining the physics use of the word "work." I said, "Imagine mowing your lawn." Everyone had a blank look until Marcus raised his hand and said "Miss, we ain't got grass." This reminded me of ice in PNG. Since all of them were familiar with shoveling snow, I was able to adapt my example to their realm of experiences. Once again I had to realize that my lack of cultural capital had implications for my science teaching. Until I was able to gain more knowledge about the types of examples my students could identify with it was difficult to get them to understand not only how physics worked but also how it played a role in their lives. As the semester continued, I learned to be more flexible and able to change plans depending on the needs of the class.

One major difference in the students at Neighborhood High compared to those in PNG was a lack of motivation to excel at school. In PNG it was considered an honor to make it through high school. In Philadelphia, many students did not see success at school as an honor. Many viewed school as a place to see their friends and get out of the house and off the street. The first time I gave a test I was surprised when Sherita and Lakisha simply wrote their names on the top and handed in a blank paper at the end of the period. When I talked with them and said I was concerned that they had failed the test, they both explained that they had not failed the test because they had not taken it. I was not prepared for their resistance toward assessment. The more I asked other teachers the more I heard that these students came from families that held the education process in suspicion. They saw school as a promoter of the oppression they had suffered for decades.

> Neighborhood High is not the easiest culture to survive in and produces students who are slow to trust and quick to react. The students protect their own and avoid or attack people they do not know. Along with all of this they are teenagers trying to find a place for themselves in a very complicated neighborhood. (Journal entry, March 2000)

Physics was an elective course so the students I was working with were considered to be above average, in that they came to school regularly and handed in assignments. Yet of the twenty, only five were actively pursuing scholarships and college applications. Four of the students were already parents who balanced parenthood, schoolwork, and jobs.

My interactions with the surrounding community were very limited. Each morning as I walked from the subway up Broad Street, I noticed most of the students crowding into McDonald's for breakfast. Yet, beyond such observations or our discussions of the last Sixers' game, I felt I had no connection with my students outside of the classroom. Parents who I contacted either seemed at a loss about what their child needed or were indifferent to my call. One mother asked my advice about how to raise her child, and I felt I did not have the tools to answer her question. I could not imagine the challenges a single mother faced raising her children in the projects. In PNG I had often wanted more contact with parents. At Neighborhood High I found that this was difficult to do, and when I was able to contact a parent, our different cultural structures often caused further contradictions in our efforts to help his or her child. I did not have the social or cultural capital to offer valid advice, and this could best be gained by experiencing, or at least understanding, the limitations that parents perceived with urban educational systems.

In most urban public schools resources are very limited. However, Neighborhood High seemed to be an exception to the rule due to an amazing department head who took great pride in maintaining excellent equipment for his staff. I had a variety of instruments to use during my physics lab. In PNG I often took the classes outside the building and drew on students' lifeworld experiences to help them understand labs. On the fourth floor we were unable to easily leave the building and the hallways were basically kept clear between classes by the school police; so we stayed in the room with the door locked and used the equipment bought over the years. This lab and its tools seemed foreign and complicated to the students and tended to make concepts even more confusing. First they had to learn to operate a piece of strange equipment and then relate their measurements back to the original concept. As a teacher I saw the importance of learning skills that would help students operate various pieces of equipment, yet I felt that alone was not motivation enough to help students learn new ideas. Again I faced the question, "How will I use this in my life outside of this classroom?" I had to get the students to focus on skills such as problem solving, following a procedure, asking questions, and applying a general concept to a specific incident. These skills would help them in my class as well as in their lives where Newton's Laws often took second place to other laws.

As the year came to an end, I had a big decision to make: Should I stay at Neighborhood High or try for a different school in the city? I was offered a job at Urban Magnet, which is considered the best school in the city. Students are accepted based on test scores and come to the school in the fifth grade. There, I would have the opportunity to teach my subject at a very high level and work with highly motivated students. After weeks of indecision and polling many of the people I knew in the city, including my department head at Neighborhood High, I decided to take the job at Urban Magnet.

The Country Club of Public Schools

Urban Magnet's motto "Dare to be excellent" is more of an expectation than a challenge. During my first year I was assigned an eighth grade homeroom, two eighth grade physical science/earth science classes and one eleventh grade physics class. My duties as an eighth grade advisor seemed overwhelming and I had no idea what was expected of me. This was compounded by the pressure placed on eighth graders to be accepted into one of the top high schools or else face the possibility of going to their neighborhood school, which was generally considered not desirable. I was unsure of many of the rules and deadlines associated with this process and had to rely on other teachers for this information. At Neighborhood High I had worked independently of other teachers, looking to them for more of a social outlet than for information. However, I found the politics of Urban Magnet to be more complicated than expected. I very quickly learned that allegiances were very difficult to find and I observed many teachers who seemed shunned by their peers. This made me anxious to find acceptance. I also felt that people withheld their judgment of my abilities as a teacher until I had proven myself knowledgeable. At Neighborhood High most people were happy just to have a warm body in the classroom. Here I had to prove that I not only knew my subject but could also navigate my way through the relationships in the building. I concentrated on building and exchanging new forms of social and cultural capital not only with my students but also with the faculty.

As far as my teaching was concerned, I was challenged in ways I had not experienced before. The students at Urban Magnet actively questioned my knowledge and my teaching. In PNG it was not culturally appropriate to directly question the teacher's capabilities. At Neighborhood High my social capital was worth more than my GPA. However, at Urban Magnet I often found students testing my knowledge, at times almost trapping me. At this early point in my career I was afraid to admit that there were things that I didn't know. I suppose that, as a student, I expected my teachers to be experts, and as a teacher I therefore expected the same of myself. I had some very long nights trying to find the answer to a calculus problem that I could not solve. I felt that I had to maintain my expertise by never claiming not to know the answer. I quickly saw this would not work at Urban Magnet. I learned to admit when I was not sure and to follow with promises to find out. I tried to show my students that teachers were people, not encyclopedias. As my students saw my willingness to learn new things, they also saw that learning was a lifelong process. This was now my third year teaching and I was starting to feel more comfortable assigning and solving new problems. When I stopped focusing on what I might mess up and started looking at the higher level learning the students were capable of, I began to truly enjoy my time in the classroom.

Part of my intimidation at this school came from certain interactions with parents. After several of these interactions, I wondered why I had once thought it would be helpful to have parental involvement. At midterms I sent home notices of students achieving Cs and below, and I realized that some parents and students were more concerned with grades than they were about learning science. Applying activity theory to these situations, I saw contradictions arise from differences in the goals of participants. I felt the learning of science was more important than a 4.0, but the parents who had to think about scholarships and college applications were focused on grades. At first, I found I did not have the tools readily available to reconcile these differences in objectives. I began calling parents to discuss plans to help students. When I called one mother, her anger surprised me. She accused me of trying to make her daughter hate science. She was an experienced teacher and seemed to know all the right things to make me feel unqualified to be in a classroom. After I assured her that I only had her daughter's best interests in mind, I tried to make the best of the situation. My own mother reminded me that for every difficult parent there were others who were wonderful with whom to work. For the rest of the semester I graded this student's papers with a sense of fear. Instead of wanting her to learn for the sake of learning, I wanted her to learn so that I would avoid her mother's wrath. Unfortunately, there had been numerous cases at Urban Magnet where the administration had not stood up to parental pressure. Therefore teachers felt that they must either compromise their values when assigning grades or face a war against people with political and economic power. As a new teacher in the school, I held little social capital with the parents and with the administration and consequently felt that I had limited resources when I was working with parents. Many parents were very supportive and wanted their students held accountable for their actions; however, I did not want to take the chance of another painful parent encounter. Therefore, instead of building positive relationships with parents, I found myself unwilling to make contacts unless absolutely necessary.

Now in my fourth year of teaching I have established better relationships with the parents of my students because I am more prepared to answer their questions. I have recognized that my lack of tools was directly related to not having anticipated the need for a guideline when working with parents. This semester when I went down my list of students who were not passing my class I called their parents with a grade sheet in front of me and a plan of action in place. Most parents were pleased to have a discussion about their student's work in my class and were eager to set up tutoring sessions if needed. As I gained support and social capital with these parents through positive interactions, fewer parents were willing to question my motives or my abilities. Likewise, as students learned I was willing to help them at various times during the day and genuinely wanted them to succeed in my class, I gained social capital with them. In addition, once my students realized that I valued their success and was committed to them, they were often motivated to help me reach my goal of having them succeed in science. This made a big difference in my relationships with students and became another tool I could use in my teaching of science.

As described in chapter 1, the student body at Urban Magnet is racially and economically diverse. Students come into the fifth grade with educational backgrounds ranging from private schools to some of the poorest elementary schools in the city. It is often disparity in resources that leads certain students to struggle and fail at Urban Magnet. Many of my students have parents with degrees from prestigious universities and salaries to match, while other students' parents never went to college and work blue-collar jobs throughout the city. I had to consider this when giving and grading assignments. Some of my students did not have access to computers. Therefore I accepted labs neatly written in black ink or allowed students to type up the lab at school. I recently asked students to create a model of a molecule from materials at home. Some students brought in very sophisticated models with beads and wires. Others brought in tennis balls connected with string. One of my students came to me the day it was due with a bag full of pennies for her electrons but no other materials. I had offered other materials at the beginning of the project but she had not come to see me. When she asked to be excused from the assignment, I brought her construction paper, markers, and glue and asked her to work on the project during her afternoon tutoring session. In this situation she was not only lacking in physical materials but also the ability to discuss her needs with me. However, it is not only the students who struggle to find resources to use for science, as I am also lacking materials to use to teach science.

Although Urban Magnet is one of the premier public high schools in the state, it is part of a resource-poor urban school district. This is when it comes in handy that I worked in a third-world country such as PNG. Students sometimes complained that the equipment was not very sophisticated, since the labs I devised often relied on only a piece of wood and a stopwatch. At the time, I felt that this was a disadvantage. However now I think that it is more important for students to learn to follow a procedure carefully, make observations based on data, and form conclusions based on their observations. These are tools they will use no matter what their future holds for them. Problem-solving techniques learned in a science class can be used in new ways in a student's day-to-day life. Still, students may not readily make these connections between school and other environments, which makes it essential that I help them see these relationships between science and their lives.

The Journey Continues

As is commonly the case in the teaching profession, the longer you are at a school, the more you get to teach in your field. Recently I have been rewarded with more physics classes. I am now teaching one section of eighth grade science, two sections of grade eleven physics, and one section of AP physics. The AP physics is a tremendous challenge in terms of content knowledge. I am using calculus that I haven't used in years. The labs we do are still hindered by a lack of equipment and once again I am forced to find alternative activities to help students learn the material. As part of my work as a teacher researcher in my classroom, a university researcher (Stacy Olitsky, chapter 16) and I have identified data resources in my eighth grade class that provide insights into how science is being taught and learned. I have also been working with four student researchers, who are involved in the process of collecting and analyzing data through interviews and videos of classes. The cogenerative dialogue that takes place between us has given me a new perspective of how my students perceive my teaching and their learning at Urban Magnet. As a teacher researcher I have had more opportunity for reflection than I have ever had before. I now see how the feedback that I obtain from students can be used to make my classroom a dynamic place where I implement changes in order to help students achieve their goals. It has also given me a chance to ask myself how my personal history has helped or hindered my effectiveness as an urban science educator.

As I look back on my career, I see many parts of teaching that are independent of where I teach. In the end people look to form relationships built on trust, respect, and fairness. Students who are trying to figure out who they are, as well as learn science along the way, look to teachers to be more than just information sources. I view teaching as a constant challenge to help a variety of people (including myself) with different backgrounds and goals to become better people. My students often quote me saying, "I don't care if you're the smartest person in this room. If you aren't a nice person you will have a hard time getting people to listen to you." I believe physics is fun and exciting, but I also believe there is more to life than physics. Teachers must develop a sense of awareness of their students' culture as well as their own. Other researchers have come to the same conclusion about urban education.

> Thus one can put forth that urban school solutions then do not lie in race, color, experience or even teaching personality. Success in educating youth . . . requires a deeply grounded understanding of the cultural influences in their lives, the strategies of action engaged across fields, the objectives behind the cultural resources, and finally, the semiotic signs that trigger their appearance within a field. (Elmesky, 2001, 168)

Teachers must find ways of bridging cultural gaps within their classrooms so that they open lines of communication with their students. This connection will enable teachers to reach their goal of teaching science and will motivate students to learn science.

In the different environments where I have worked, parents have played a variety of roles. Before I began teaching, I had imagined that a greater contribution from family could only bring about positive change in schools. Now I see that the relationship between a teacher and a student's family can be just as destructive as positive if it is not carefully cultivated. The cultural differences between teachers and students in classrooms become even more complicated outside of the class, where parents in a community may follow different rules and possess goals that may conflict with my or the students' objectives. Many times a parent's frustration toward me is a result of a miscommunication with their child. Through these experiences, I have learned the importance of establishing guidelines that enable me to work with parents, not despite them. These rules are based on the fact that there is no single person to be blamed or acknowledged for a student's progress, but that students, teachers, parents, and other community members influence students and must be given equal consideration when addressing problems. My contact with families has also taught me to learn more about my students' lives outside of the classroom in order to better understand the expectations of their parents.

Last summer I visited a private school with a high-tech lab that must have cost a small fortune. I had not had the opportunity to work in such an environment since I was in college. I do not feel that this has been a disadvantage for my students in their learning or for my development as a teacher. The resources that are important in the learning of science are not always what can be ordered in a catalog. Instead they tend to be resources such as social and cultural capital that can be used to gain students' attention and respect. I want to continue to incorporate student experiences in my science class and teach them to utilize available tools. I realize that not all of my students will major in physics; however, I feel that critical thinking, problem solving, and cooperative work are skills that will be useful in any field. Especially in an urban setting, improvisation is a beneficial tool for students to learn, considering few schools have adequate or appropriate resources for students and teachers to use.

It seems that the more I teach, the more I learn about my subject and myself. Because of my recent urban teaching experiences, I have become more interested in how teachers in urban settings draw on their own history as a tool to reach their objectives. Teachers in the inner city must distinguish between parts of their lives that have prepared them for this work and parts that cannot be applied in this challenging environment. They must examine closely their own culture as well as their students' and find a peace between the two without compromising either's integrity. This approach could help teachers deal with the variety of challenges that will arise from teaching in a community with which they may have very little in common. It is not a coincidence that cities like Philadelphia, New York City, Washington D.C., Miami, and Los Angeles recruit returned Peace Corps volunteers to teach in their school districts, since they need people who are flexible, adaptable, and able to work in resource-deprived areas. To be effective in these places, teachers must find their work rewarding because of (and not simply in spite of) the rich diversity of the students and the challenges unique to the urban setting.

Chapter 16

Social and Cultural Capital in Science Teaching: Relating Practice and Reflection

Stacy Olitsky

School science can be an alienating experience for students, particularly for those in low-income, urban areas, due to factors such as the use of unfamiliar communication patterns that may favor students with middle-class backgrounds (Lemke, 1990) and a lack of connection between school science and students' lives (Tobin, Seiler, and Walls, 1999). Teachers face the challenge of negotiating between the expectation of designing lessons to adhere to a highly structured science curriculum, especially where standards-based reforms have been implemented, and the goal of designing lessons to be congruent with students' cultural backgrounds and interests. This process is often a difficult, solitary endeavor for teachers in urban schools. In this chapter, I explore how a teacher's involvement in collaborative, reflective research grounded in the use of social theory helped her better understand the challenges she faced teaching students from different cultural backgrounds, thereby providing insights that improved teaching and learning in her classroom.

This study centers on the classroom of Linda Loman, the author of chapter 15, who teaches science at Urban Magnet, a diverse urban magnet school. Linda and I worked as coresearchers in her classroom over the 2001-2002 school year.

Through collaborative, reflective research we examined her teaching goals, which included helping her students to achieve in science, to perceive science as relevant to their lives, and to identify themselves as people who can do science. In addition to analyzing data from her classroom, we engaged in reflection and cogenerative dialogue (Roth, Tobin, Zimmermann, Bryant, and Davis, 2002) with a research team composed of Linda, four student researchers, and me (a university-based researcher). All four of the student researchers are female and came to Urban Magnet from neighborhood elementary schools throughout the city. Ashley, Emani, and Monique are African American, and Lisa's father is white and her mother is African American. As told in her autobiography, Linda is white, in her late twenties, and came to Philadelphia from New Mexico.

Rodgers (2002) writes about Dewey's (1916/1944) idea of reflection as being a rigorous, structured process that involves detailed description, analysis, and development of theories that can be applied to understanding experience. As part of structured reflection, Linda wrote an autobiographical paper (see Loman, chapter 15) in which she traced her development as a teacher, through her experiences in Papua New Guinea, New Mexico, and a neighborhood school and magnet school in Philadelphia. In applying social theory to her experiences, Linda identified several themes relevant to how she redefined her role as she negotiated the professional expectations that she felt came with being a science teacher and her changed expectations based on her growing awareness of the importance of relating to her students. The themes she chose included the material resources for teaching science, family involvement, cultural capital, and social capital. This companion chapter examines both the intentional and the habitual aspects of Linda's teaching with respect to two of the themes laid out in her autobiography, social and cultural capital, in the enacted science curricula.

Our co-constructed research questions included the following: How did Linda use the concepts of social capital and cultural capital to understand her past teaching experiences, to formulate her ideas about what it means to be an effective teacher, and to guide current practices? What contradictions did Linda face in incorporating students' cultural capital into the classroom and building social capital with the students, and how did she address these contradictions? How did the interpretations of the student researchers regarding the role of social and cultural capital differ from the interpretations held by Linda, and what insight can these differences provide on Linda's habitual and deliberate practices that students perceive as helpful? What are the benefits and limitations of the collaborative reflective research process for both improving the science classroom and enhancing Linda's development as a teacher?

This chapter describes how the reflection and analysis conducted by the research team on the role of social and cultural capital increased Linda's awareness of the impacts of her strategies in the classroom, helped her formulate plans for developing and emphasizing successful strategies, and increased mutual understanding between participants. Over time, Linda obtained a more detailed view of how social and cultural capital related to each other in her classroom, as she saw how incorporating what she viewed as students' cultural capital not only helped her to work toward her goal of raising student interest in science, but also helped her garner greater social capital with the students. In addition, she became more aware of some of the barriers to her efforts to make science more connected with the students' lives. The results of this study suggest that a structured, collaborative, reflective process involving the application of social theory to understanding classroom events can be a useful tool for improving teaching and learning in urban classrooms, particularly in science where students often feel alienated from the subject.

Reflection and Autobiography

Reflective teaching has been viewed as an important goal of many teacher education programs (e.g., Allen and Casbergue, 1997). Rodgers (2002) describes John Dewey's (1916/1944) conception of reflective practice as involving teachers' reconstruction of experiences in order to obtain meaning from them. She writes that for a novice, time between thought and action is essential for this process to take place. However, in the eventful nature of teaching, there is not much time for reflection (Roth and Tobin, 2001), as teachers need to respond immediately to the many situations that arise during instruction. Roth and Tobin describe how teachers' actions often do not emerge from deliberate, conscious thought, but instead from the teachers' habitus, or dispositions, which includes both perceptions and patterned actions that structure how individuals act in particular situations and settings (e.g., Bourdieu, 1977). While habitus develops through socialization (Bourdieu, 1977), Roth and Tobin (2001) describe how it can continually change with experience. As teachers spend more time in classrooms, they learn how to approach new situations and develop different ways of acting in the world. A more experienced teacher will have a greater range of actions to draw on in any particular situation, giving the teacher more room to maneuver (Roth, Lawless and Masciotra, 2001) and helping him or her to be more effective in responding to the challenges of teaching.

A view of teaching based largely in nonconscious dispositions rather than intentional behavior still suggests a crucial role for reflection. Donnelly (1999) writes that people are more deliberative regarding their actions when they face situations in which their current set of strategies is ineffective. I use the term strategy to refer to both conscious and nonconscious actions that emerge in particular situations. When facing unfamiliar challenges in eventful classrooms, teachers are unable to act out of habit when their current strategies fail (Roth and Tobin, 2001). Instead, at those moments they become more conscious of the situation, reflect, and develop new approaches. Over time, these newer, more effective approaches become a part of their teaching habitus. Thus, reflection in response to difficult situations, in which habitus has broken down, can be an important way of acquiring new ways of being in the classroom. Rodgers (2002) writes that even for expert teachers, time to reflect is useful, since they can become more aware of what they already know implicitly.

Through constructing her autobiographical narrative, Linda was able to reflect on how her goals and teaching strategies had changed as a result of encountering situations where her previous ways of being in the classroom were not effective. Schiffrin (1996) discusses how we use narrative to "represent ourselves against a backdrop of cultural expectations about a typical course of action; our identities as social beings emerge as we construct ourselves in relation to social and cultural expectations" (170). The social and cultural expectations that teachers confront in their experiences can also be looked at as types of schemas (Sewell, 1999). Sewell bases his conception of schemas on Giddens's notion of "rules," which encompasses "not only the array of binary oppositions that make up a given society's fundamental tools of thought, but also the various conventions, recipes, scenarios, principles of action, and habits of speech and gesture built up with these fundamental tools" (8). Sewell writes that schemas, along with resources, constitute the structure that both shapes and is shaped by people's practices. People are constrained by structure, but can act agentically in transforming it as they mobilize resources in creative ways or reinterpret schemas in order to further their own goals.

Linda's autobiography provides a historical perspective of how she responded to barriers in her efforts to be a successful science teacher, how she reconstructed her schema related to what it meant to be a good teacher based on her experiences, and how she engaged in different teaching strategies in accordance with her reformulated schema. Linda explains how she initially had the idea that she could "simply walk into a classroom and disseminate information," suggesting that she felt that her primary goal was to give students knowledge of science. She describes how her thoughts changed over time when she had to confront cultural differences in the classroom. As she responded to challenges posed by the tension between needing to teach what she considered to be the culture of canonical science and needing to teach in ways that would be understandable to students from cultural backgrounds that differed from her own, she expanded her expectations of the teacher's role to include knowing and respecting students' cultures. In chapter 15 of this volume Linda concludes, "They [teachers] must examine closely their culture as well as their students' and find a peace between the two without compromising either's integrity. This will help teachers deal with a variety of challenges that will arise from a community with which they may have very little in common" (279). The challenges that Linda describes may be similar to those that other science teachers face in a time of increased standardization, where there may be disconnects between what and how teachers are expected to teach and what is understandable and relevant to students.

Rodgers (2002) supports her view of the benefits of collaborative reflection for teachers by drawing on the work of Dewey (1916/1944), who wrote about the importance of communicating one's experiences to others. Dinkelman (2000) writes that reflective teaching can develop more easily if a new teacher works with a cooperating teacher who values reflection, or takes courses that teach reflective teaching explicitly. However, even if reflection is encouraged by mentors, a teacher is typically still limited to her own, or other adults', reconstructions of events, since students do not usually play a large role in the process. It was our belief that reflection could be enriched by students' voices, and this belief influenced our approach of combining teacher reflection on past and present practices with students' interpretations of events. Thus we expanded the collaboration in reflection to include Linda, students, and me.

Applying Theory to Practice

In interpreting her experiences, Linda drew on activity theory (Engeström, 1999) and the work of Bourdieu (e.g., 1986) on the role of different forms of capital, such as social, cultural, symbolic, and economic capital in social life. In the process of writing her autobiography, Linda read about these ideas, discussed them with her research team, and reinterpreted the concepts in relation to her own experiences.

Bourdieu (1986) described how forms of capital are exchangeable and have a powerful influence on people's social positions and on their ability to attain their goals. In our research, we use the term social capital to refer to the resources generated by relationships. Cultural capital includes the knowledge, skills, dispositions, uses of language, and other conscious and nonconscious attributes that can either help or hinder people in achieving their goals. Expanding on Bourdieu's work, Lamont and Lareau (1998) have conceptualized cultural capital as a basis of exclusion, which can be helpful for thinking about how unfamiliar language, content, or approaches to argumentation (Lemke, 1990) can sometimes exclude students from science classes. Cultural capital is most often thought of in relation to the exclusion of those with less power from domains controlled by those with more power. In this sense, students are a group that is subject to exclusion in classrooms. However, Linda also used the term cultural capital to refer to instances where she felt excluded as she endeavored to interact with her students and had to acquire new cultural capital in order to reduce conflicts with students, parents, or other staff members.

An activity-theoretical perspective (Engeström, 1999) can also provide insight into how social and cultural capital influences people's ability to achieve their goals in a particular setting. Engeström views activity as a collective endeavor, in which the path between the subject and the goal is mediated by components such as rules, resources, and the division of labor. Students' and teachers' cultural and social capital can be considered resources that participants can use to achieve their goals within a classroom activity. However, contradictions, inconsistencies, or conflicts between and within these components can interfere with actors being able to meet their goals. For example, low student achievement and confidence in science could be seen as partially resulting from a contradiction between the resources that the teacher expects students to have for the learning of science, and the resources students actually bring to the classroom. Studies have shown that in urban classrooms, students' cultural capital is sometimes not valued, which can impede their science learning (e.g., Seiler, 2002).

Applying activity theory also provides ideas on several possible ways to bring about change. One manner of change in a system is when participants work together to eliminate contradictions and strengthen patterns of coherence (Engeström, 1999) thus strengthening the achievement of collective and individual goals. In Linda's classroom, the insights generated through joint reflection with students enabled Linda to see how incorporating students' cultural capital in the classroom not only helped students feel less distant from school science, but also helped her to build social capital with the students and helped them to earn it with peers. In addition, Linda gained insight into the importance of students earning social capital so that they would feel comfortable being active participants in classroom discussions and activities, and how she could best facilitate this process. Her growing understandings of how these forms of capital interacted in her classroom became an important resource to improve her teaching.

Who, How, and Why

Our research takes place in an eighth grade science class at Urban Magnet. The approach we employed was influenced by Guba and Lincoln (1989), who developed criteria for authenticity that includes working with participants toward positive change and increasing their understanding of each other's perspectives. Toward this end, we conducted the research as a team composed of Linda Loman (a teacher researcher and author of chapter 15), four student researchers from the class, and me (a university researcher). Data were collected in the form of field notes, student work, interviews of and by students, and video and audiotapes of class sessions, interviews, and peer guidance sessions. We coordinated and analyzed multiple data sources in order to generate a more valid picture of the events of this classroom.

The student researchers' roles were to participate in joint discussions of classroom events, analysis of videotapes of class sessions, analysis of transcriptions of class sessions during biweekly research meetings throughout the school year and daily during the following summer, and to interview their peers about science learning. In selecting student researchers, we identified four students who were different from each other in terms of their academic achievement, as measured by grades, and their interests in science. The students offered diverse perspectives and interpretations of classroom events, which led to many lively discussions.

As the university researcher, I acted as a participant observer in the classroom, sometimes coteaching lessons, holding weekly supplementary science sessions during lunch, and meeting with Linda after class or during her lunch period to reflect on the events occurring in the classroom. During the science lunch sessions, I had the opportunity to interact informally with students and learn more about their experiences with schooling and science. Linda described one aspect of my role as being an adult for students to approach who was not there to evaluate their performance, but who was simply there to help them in their class.

I also held somewhat of a facilitator role during the cogenerative dialogues/research meetings. While all members of the research team raised issues and questions, I began most meetings with topics and video clips that could serve as initial foci for discussion. These topics were most often prepared in collaboration with Linda or based on conversations with the students. During the dialogues, I sometimes posed questions and invited participants to clarify their statements in order to insure that participants understood each other's ideas and that conflicting opinions on various issues were discussed rather than de-emphasized in favor of consensus. In addition, I introduced theoretical frames to the dialogues and helped students develop their interviewing and research skills.

Through our collaborative research approach, we attempted to avoid a traditional researcher/researched relationship that potentially exploits those being researched. While we recognized that divisions in power would persist regardless of the methods that we employed, we felt that we could reduce them by conducting a study where all participants would be involved with formulating goals, evaluating data, and discussing results. An emphasis was placed on discussion and reflection in which student researchers, in addition to the adult researchers, would be involved not only in providing feedback on the science class, but in using theoretical lenses to analyze classroom events. It was our hope that by having all research participants learn and apply social theory, the voices of the adult researchers would be less privileged than would otherwise be the case. The students had the opportunity to be critical of the theoretical lenses, apply them in creative ways, and dispute the adult researchers' interpretations. While Linda and I recognized that the students' understandings of the theory would differ substantially because of their age and experience, the students were able to interpret ideas based on their own experiences and apply them to their own analysis of classroom events. Our research meetings were transcribed and served as a source of data, generating insights on the multiple perspectives held by participants. Having both Linda and the student researchers analyze the same events led us to collective insights that would otherwise not have been possible. The following example shows how this process contributed to Linda's awareness of both the benefits and limitations of her habitual and deliberate actions involving the use of social and cultural capital to accomplish her goals in the classroom.

Transposition of Social and Cultural Capital

A recurring theme in Linda's autobiography was how cultural differences between her and the students affected her success in teaching. These differences in practices, dispositions, and knowledge between her and the students represent differences in cultural capital. Linda's success in addressing cultural differences in order to attain her teaching goals was influenced by both her own efforts and school structures, such as rules, resources, and curriculum. Over time, Linda's construction of a schema for herself as a teacher changed as she began to recognize the importance of knowledge of her students' cultures in teaching science.

Linda relates how she had difficulty teaching physics in Papua New Guinea because of differences between the examples and materials associated with canonical science and the knowledge that the students possessed. On one occasion she was able to incorporate student knowledge by using canoes rather than lab equipment to demonstrate Newton's laws. In telling this story, she describes how she learned about the students' cultural capital, in the form of knowledge of how canoes behave in the water, in order to share her cultural capital, in the form of canonical science knowledge. Because she had the freedom to adapt the curriculum in this particular school, and because the students and teacher both held goals relating to the teaching and learning of science, she was successful in both respecting students' culture and in teaching them science.

At other times, Linda found that cultural differences were not as easily bridged, such as when she faced the Papua New Guinea students' beliefs in spirits as an excuse for absenteeism. Linda's narrative describes how in these cases, she could not address the contradictions simply by changing how she taught physics. Instead, she needed to develop a respect for their culture that might then allow her to gain social capital. She describes her experiences in Papua New Guinea: "In this community, I learned to follow many of the social structures so that I would be accepted by my students as someone who respected their culture while maintaining my own culture and teaching them about some of my values, rules, and of course, physics." Even if she did not feel that she could directly incorporate certain aspects of the students' cultural capital into her life and her science teaching, she felt that understanding it was crucial to obtaining social capital, thereby meeting her teaching goals.

Central to Linda's developing schema of what it means to be a good teacher is to have an understanding of students' culture. She quotes Elmesky (2001), who wrote, "Success in educating youth . . . requires a deeply grounded understanding of the cultural influences in their lives" (168). However, the diversity of the students in Urban Magnet made the challenge of "grounded understanding of cultural influences" somewhat more difficult since, other than influences such as popular media culture and high academic performance relative to other students in Philadelphia, there were few similarities between all her students. In addition, unlike Papua New Guinea, Linda's interactions with the students were usually limited to class time and informal encounters in the hallways. Yet, if the structures of Urban Magnet and the characteristics of the students did not afford Linda sufficient opportunities to acquire deeply grounded understandings, how did she approach her goals of considering cultural and social capital in her science teaching?

make them feel any closer to science, because in the words of Ashley, "It's the same science. She could use stick figures, it doesn't matter. It may help us to remember it though." Ashley's statement suggests that she sees the boundaries between science and other forms of cultural capital as more rigid than perhaps Linda is intending. Other students' statements supported this view. Lisa discussed Linda's use of sports examples, "It's not really science, its just stuff to help us learn it." All of the student researchers did say that using examples from sports helped them to visualize and remember the ideas better.

The comments by the students on this vignette, suggesting that they view the boundaries of science as somewhat rigid, can be considered together with other statements the students have made in the past on the role of science in their lives. The student researchers seem to have a conception of science as cultural capital that can facilitate their ability to meet some of their future goals. For example, Ashley wants to be an obstetrician, and she knows that she needs to learn science, partially from hearing about the experiences of her sister who is in college. In their discussions, the student researchers often depict science as a fixed body of knowledge that mediates their entry into certain domains, such as high school, college, or medical school. Their statements indicate that they see Linda's inclusion of familiar examples as helpful for them to access this potentially important cultural capital through improving their memory, but they do not seem to perceive that the boundaries of science are changed.

Developing an identity as someone who can do science does not seem to be a part of Ashley's conscious articulation of goals, as it is for Linda. However, the inclusion of such cultural capital may help them to see science as more accessible, even if they are not aware of it. Still, their perspectives may point to schema and aspects of the school activity system that limit Linda from achieving some of her goals. These students are in an academic magnet school, and many of them are very aware that their futures are significantly affected by their grades, performance on standardized tests, and whether their coursework is considered rigorous enough to gain them eventual college admission. In addition, some of these students will not be accepted into the more prestigious high school if they do not perform well. Therefore, it is not surprising that they speak of science as fixed, potentially exclusionary cultural capital regardless of the way it is taught. A view of science as rigid and strongly bounded, supported by schools' testing and admission policies, poses contradictions that interfere with Linda's goal to use students' cultural capital in order to help them feel closer to science.

It seems salient that, in discussing what they liked about Linda, the students emphasized positive interpersonal relations. It is not the use of sports itself that student researchers seem to value. Rather, Linda's efforts to include their cultural capital along with her joking show she is trying to make the class and their school experience better for them. Linda's goal in using sports knowledge in the class may be to have students feel that science is relevant to their interests, but to the students, who still conceive of science as rigid, such attempts at least earn Linda social capital that contributes to her effectiveness as a teacher.

Reflecting on a day's events in isolation, Linda might have been able to identify particular strategies as successful based on whether she felt she received positive responses from the students in class. However, reflecting collaboratively with the student researchers allowed her to gain a different perspective on how students perceived her actions, why students believed that particular strategies may be successful, and how contradictions within the activity system may limit their effectiveness. The students' perspectives, combined with her own reflections, became resources for her in figuring out which strategies to emphasize and develop.

The Students Earn Social Capital with Peers

While most of Linda's autobiography concerns her own experiences with differences in cultural capital and her attempts to build social capital, in many research meetings she addressed the issue that as a teacher, she also needs to be concerned with students' access to social capital among their peers. Due to the nature of the school culture at Urban Magnet, some students may lose social capital if they do not have the requisite background to answer certain types of questions correctly or perform highly on tests. Losing social capital in this manner can negatively affect their school performance and, ultimately, their feelings about science. She therefore tries to provide opportunities for all students, regardless of background in science, to participate successfully. Her assignments include many creative tasks, such as skits, art projects, and movies. During class discussions, she asks questions that she believes students will answer correctly, including types of questions that allow students to use skills and knowledge other than those that are traditionally associated with science. She often elicits personal experiences from students or invites them to invent mnemonic devices to help remember new material. One of her goals is to enable all students to earn social capital with other students through science participation, thereby fostering positive identities associated with science learning.

The following transcript was identified by Linda as an example of how she enacted a strategy of providing opportunities for students to use their cultural capital in order to participate in science. In this vignette, two students, who are often quiet in class, volunteer to provide ways for other students to remember the phases of the moon.

Rick   When you wax on, you put something on, when you wax off you're taking it off.
Ms. L   Yeah, that's a good way using the karate kid.
Ashley   Yeah, that's the karate kid.
Ms. L   Yeah, that's the karate kid, that's the karate kid, ummmmm, James.
James   If the moon is dark when you're about to wax it and when you wax something it gets shiny, and when there's a full moon it's shiny and bright, so when you wax it its shiny.
Ms. L   So when you're waxing, you're trying to see a better shine. Can anybody else think of a waxing and waning analogy? OK, so waxing goes from nothing to the full moon and then waning goes from this bright big huge thing to nothing.

When examining the videotapes of the classroom, our team recorded differences in frequency of students' oral participation, and observed that the students who generally tended not to volunteer to answer factual questions were often the first ones to volunteer to talk about a personal experience, read their story of the constellations, or recite their strategies for remembering the required formulas or lists. Emani, one of the student researchers, discussed questions involving mnemonic devices, saying, "They can answer them and be right. It's more fun that way . . . People don't like to be put in the spotlight and others say that they are stupid." Ashley said, "People like to talk if they do not think they are going to be wrong. Otherwise, they will just let Peter [a high-achieving student] answer." Lisa explained that some students avoid "some types of factual questions . . . it's the same kind of knowledge in any class, it's that knowledge the teachers really want you to get." The students agreed with Linda that they earn symbolic capital by answering questions, but Monique said that easy, factual questions earn less capital than questions that allow them to bring in outside knowledge, since "everyone knows if a question is easy."

The students' discussions about why they think students volunteer more for some types of questions than others suggests they may avoid "factual" questions, despite their seemingly clear-cut nature, because they risk losing symbolic capital by giving an incorrect answer, and they experience a contradiction between the goal of participation and the goal of attaining status among their peers. By providing questions and assignments that incorporate students' creativity and cultural capital, Linda establishes an atmosphere where students who feel alienated from traditional classroom talk can make a contribution to classroom science and gain status. Based on observations of the class sessions, Linda's strategy seems to be effective in helping more students feel comfortable participating in whole-class science talk, thereby strengthening patterns of coherence in her classroom.

However, many of the student researchers' comments indicate that the students perceive a separation between "science" and "things to make science more interesting" or "things to make us remember." If students believe that only the factual questions are categorized as "real science," some students may not view their own contributions as important to constructing science knowledge and may, therefore, not view themselves as people who are successful science learners. Moreover, structures such as standardized testing that may have contributed to students' perceptions of the boundaries of "real science" may interfere with Linda's efforts toward her goal of facilitating a more inclusive view of science over the long term. Joint reflection has increased all of the participants' awareness of this potential barrier.

Linda intentionally uses other strategies to enable students to gain social and symbolic capital, such as when she encourages Emani and James to use humor in their responses to questions about science, since humor seems to be one of the methods that these two students use to earn status with their peers. During a research meeting, Emani answered a question regarding how she exercised agency in a video clip from the classroom, writing, "I got to make people laugh and I answered a question." By encouraging students to use humor in classroom discussions, Linda helps establish an environment where students can participate in science class, earn symbolic capital, and earn social capital with peers. It seems likely that this approach will help Linda reach her goal of increasing student engagement with science over the long term, as students may be more willing to participate on a deeper level in the future than if their efforts at participation afforded them little gain in social status.

While Linda was already aware of students' needs to earn social and symbolic capital before her reflection and research with the team of students, in the beginning of the year she had spoken mainly about giving students easy questions that they can answer correctly. However, the student researchers' comments demonstrate that in their opinion, such "factual" questions are risky. Instead, students are better able to earn social and symbolic capital when they can demonstrate creativity and display outside knowledge while still engaging in science discussions or projects. Through collaborative research, Linda formed a more detailed view of how various instructional strategies were perceived by students, and how strategies that emphasize the use of students' outside knowledge, which she refers to as cultural capital, can help students earn social and symbolic capital with peers.

Linda Earns Capital through Advising

Linda's autobiography indicates that she developed an expanded view of the teacher's role to include being a supportive adult in the students' lives. Serving as the eighth grade students' advisor not only helps her achieve this goal of supporting students, but also improves her science teaching because it provides her with additional opportunities to earn social capital with the students. The student researchers have discussed how one of the reasons why students listen to Linda, yet not to some of their other teachers, is because Linda serves as their advisor and speaks to them every morning. When she interacts with them informally, Linda shows the students respect and caring, thereby earning social and symbolic capital.

While Linda is able to act agentically in establishing an environment of care, she encounters structural constraints, as well as structures that facilitate building social capital. While she would like to know the cultures of her students, it is difficult because of the diversity of her classroom. Instead, Linda tries to understand them the best that she can, and compensates for her lack of opportunity to gain such knowledge by demonstrating concern and respect. The advisory period serves as a structure that facilitates her success with the students, because she can talk to the students informally. Ashley explained, "I think one of the reasons why students will learn from Ms. Loman is that she is our advisor, and she talks to us in the mornings. Also we're her only eighth grade class. With some other teachers, they say, 'did I talk with you about this or was that the other group?' You're like . . . she doesn't even know us." Linda's schedule gives her extra opportunities to gain social capital with the eighth grade students. However, these same structures constrain Linda from gaining social capital with her other classes. While she has opportunities to interact with her eleventh grade students informally in the halls, the large number of students she teaches and her limited time with them may pose contradictions that interfere with her goal of supporting all her students.

More Inclusive Science

One important goal for Linda, as stated both in her autobiography and in cogenerative dialogues, is for all of her students to identify themselves as people who can do science. She writes in her autobiography, "I saw my role as someone who could show these students that physics was not so hard." Through the process of reflective research, Linda has become more aware of and increased the use of strategies that further this goal, in particular using culturally relevant examples and providing opportunities for students to incorporate their interests into the classroom.

Linda's use of students' interests and knowledge was intended to reduce the sense of exclusion that students with diverse forms of cultural capital may experience in a science classroom where equipment, language, and methods of argument may be unfamiliar. Linda also wanted her actions to encourage students to perceive the boundaries of science as porous. Through reflection with the student researchers we learned that in spite of Linda's efforts the students seem to separate a body of knowledge that they call "science" from the expression of their own cultural capital, even when it is used in science class. While Linda has not yet been able to bring her students to think of *science* as more inclusive, they do perceive *their science classroom* as inclusive. By encouraging students to use their own cultural capital, Linda earns social capital, facilitates students earning social capital with peers, and thereby increases students' comfort participating in class discussions and activities. It is also important to note that just as Linda may not be conscious of all of the ways that her teaching is continually developing in order to help her accomplish her goals, the students may not be aware that their confidence in themselves as science students is increasing and their schema related to science are changing. Linda may be fostering conditions that will facilitate these students becoming more active participants in school science over the long term.

As demonstrated in her autobiography, Linda's ideas of the role of science teacher changed as she responded to various circumstances. She came to view a teacher's role as including learning about students' cultural capital and using this knowledge to make science seem less distant from the students and to build social capital. At the end of her autobiography, she writes,

> I view teaching as a constant challenge to help a variety of people (including myself) with different backgrounds and goals to become better people . . . I believe physics is fun and exciting, but I also believe there is more to life than physics. Teachers must develop a sense of awareness of their students' culture as well as their own. (277-278)

By reflecting on her current teaching practices in the process of research, Linda was able to gain more insight into the relationship between social and cultural capital, which was helpful in planning to accentuate current effective strategies and develop new ones. Through collaboration and cogenerative dialogue, Linda saw how bringing in students' knowledge and respecting their cultures did not just facilitate their participation in science activities, but it also demonstrated her caring for her students and helped students to earn status with each other. Often, Linda's analysis of social and cultural capital seemed to be focused on science discourse and activities, while the students' analyses emphasized relationships. Because Linda believes that the building of social capital in the classroom is likely to lead to increased engagement in the collective activity of science learning, her improved understanding of the impact of bringing in students' cultural capital on earning both social and symbolic capital can be helpful as she develops her teaching strategies.

The analysis of Linda's classroom also highlighted some of the structures that hindered her in reaching her goals of knowing students' cultures and helping students feel that science was not distant from their lives. The structures that she had to contend with at Urban Magnet included her inability to gain substantial knowledge of the students' home cultures given student diversity and lack of time, and student views of a strongly bounded science, reinforced by a competitive school environment. While several of the student researchers described the familiar examples and projects that Linda consciously tried to use as important in their science learning, they did not speak of them as opportunities for incorporating their culture into science.

This study suggests that autobiography, reflection, application of social theory, and collaborative research with students can be powerful in generating insights into improving teaching and learning in an urban school. While Linda would have made efforts to improve her teaching without engaging in research, the research process facilitated Linda's reflection using social theory as a lens and generated a better understanding of her students' perspectives on classroom events. Autobiographical narrative helped Linda to become more aware of different expectations and pressures that she faced and to clarify how she had revised her schema of what it means to be a good teacher. Through joint examination of her current classroom practices with students, Linda became more aware of strategies that were helpful to her in meeting her teaching goals. Structured reflection combined with cogenerative dialogue with students may be a helpful approach for other science teachers in urban areas as they face the challenge of negotiating the different aims of preparing students for tests and designing activities that address students' needs, interests, and experiences. In addition, the specific findings of this study regarding the importance of incorporating students' cultural capital and providing opportunities for the building of social capital could be relevant to teachers of diverse groups of students looking for ways to establish a classroom environment where students are more comfortable engaging with school science.

Editors' Perspectives

In urban schools, when social, cultural, and class differences between the teacher and students are often salient, it is often difficult for teachers to "see" the students' capital. It is one thing to exhort teachers to begin with what the students know and teach accordingly and quite another to actually do it. Too often teachers look and do not see the capital but rather the deficits and the teaching that occurs is as if students are starting from scratch. So, an initial challenge is to identify the capital of students and create a curriculum to enable them to use what they know and can do to build fluency in science. A good place to start is to interact with students about their interests and lives outside of school. A barometer for the success of these interactions is emotional energy. Teachers should maximize positive and minimize negative emotional energy to create a climate in which social networks can develop and symbolic capital can be built, especially respect. If interactions produce positive emotional energy, then proceed; however, if negative emotional energy is produced, try a different tack. The creation of social and symbolic capital is obviously connected closely with the extent to which interactions are or are not successful in attaining their outcomes (i.e., producing and reproducing the culture of science).

Chapter 17

Transforming the Future while Learning from the Past

Kenneth Tobin, Rowhea Elmesky, and Gale Seiler

Agency | Structure Relationships in Science Education

Considering the unsatisfactory state in which urban science education currently exists, and has persisted over the past few decades, many would argue that there is little that we can learn from studying what has occurred and is occurring. However we have chosen, in this book, not to take such a deficit perspective. Rather, in the chapters presented, we look backward to move forward by recognizing and trying to deeply understand how and why urban youth learn science and build identities around science *in spite of*, rather than *because of* how we teach. Certainly, the authors have shared many rich images of urban youth (many of whom have already been pronounced as academic and, in some ways, societal failures) as avid and deep learners of science. Moreover, in many of the chapters sharing instances of substantive learning, there seems to emerge, as an interesting pattern, a dynamic flow and interplay between the classroom structure (i.e., the practices of teachers and students; the resources available to access and appropriate science curriculum; and the schema) and student agency in the field of the science classroom. Accordingly, we regard the structure | agency dialectic as central to looking back and moving forward in our understandings of urban science education and as the point of leverage on a fulcrum of change in science classrooms in inner-city schools. This chapter combines discussions around structure and agency throughout the book to summarize what we have learned, establish where we currently are, and move with sure steps to a future for urban science education paved by transformation, empowerment, and a clear definition of the purpose of science education in this new millennium.

Physical and Organizational Structures

Traditionally, urban schools have been characterized in terms of a relative shortage of economic capital being available to support material and human resources to the same extent as suburban counterparts. In this study we have seen ample evidence of a shortage of material resources, and this has been exacerbated by factors within the schools that create inequities with regard to accessing those resources. Moreover, we have found that a critical factor determining the ways in which resources are accessed and appropriated is the organization of some schools into Small Learning Communities (SLCs) or schools-within-a-school, as well as whether or not a science department exists. At City High, for example, the lack of a science department was associated with a lack of an alternative organization to ensure that someone had the role of ordering and maintaining equipment and supplies for science education, repairing broken equipment, storing and disseminating equipment and materials, and safely disposing of wastes. Instead, it was left for individual teachers to decide on their needs and to compete for the budget to obtain them. Teachers who were aware of their needs for particular materials and supplies often procured them and stored them for their own convenience, making it virtually impossible for others to access them, even though the resources may not have been needed on a continuous basis. Similarly, when supplies were depleted, it was left to the teacher whose class used them, to initiate an order to replace the supplies. This task often was overlooked or was forgotten due to teachers' other responsibilities. Thus, the absence of a science department and lack of a formal position of head of department resulted in an organizational framework that could not assign roles and responsibilities for managing the material and human resources for science education in the school. Accordingly, issues associated with the roster and teaching assignments, the use of specialized rooms, and even the nature of the science curriculum were left to be decided within an alternative organizational structure, that is, within Small Learning Communities (SLCs).

Linda Loman's research on her teaching at Neighborhood High also alludes to the importance of leadership in promoting science education (chapter 15). Her experiences illuminate the ways in which teacher and student agency can be afforded through a department head who was effective in securing budget, maintaining equipment in a functioning state and having a storage system that allowed all teachers to access the materials they needed to support their teaching. At that time, the presence of a head of department was most unusual in Philadelphia's schools since schools organized around a department structure were being moved toward the SLC model. In contrast to Neighborhood High, when Loman began teaching at Urban Magnet she soon experienced many of the problems experienced by Sonya Martin at the same school (chapter 13) and other teacher researchers involved in the study. For example, like Tracey Otieno (chapter 11), Loman was given a floating schedule, where she had to move from room to room and take with her the texts, equipment, and materials she needed for each class. Additionally, like Martin, she could not access some equipment, which was hoarded by her experienced colleagues, and she was excluded from social networks among senior teachers.

In chapter 4, Roth drew attention to the school as an organization and identified contradictions that arise from the need for participants to participate in multiple overlapping fields. Central to the resolution of such contradictions is their identification, and the building of consensus around what changes are needed to resolve the problems. Roth's research illuminates how the practices of individuals can change the structure of a field and thereby the agency of all participants in that field. Thus, he reveals that the actions of individuals can make a positive difference for the individual as well as the collective, and it is not inevitable that oppression is always reproduced through social activity. His chapter, then, raises questions as to what role teachers can have in changing the structure of science education.

When Teachers Change the Structure

Across the diverse physically and organizationally structured urban schools in which we have undertaken the research described in this book, we have experienced numerous examples of teachers exercising agency to enact practices that were conducive to high quality science education. For example, Martin (chapter 13) and Sarah-Kate LaVan (chapter 14) emphasize the manner in which a teacher's agency can be truncated by the practices of others in a school. Specifically, at Urban Magnet, the practices of school administrators and senior colleagues restricted the resources available to support Martin's teaching of science. However, her agency in accessing available materials and her uses of LaVan and her students as resources in coplanning and coteaching afforded highly productive learning environments. Certainly, numerous examples from these chapters make it clear that the teaching of science need not be impoverished just because equipment and supplies are limited and the school is fiscally strapped, and the following examples from City High further illustrate this essential finding.

At City High, when Cristobal Carambo was reassigned from an SLC in the basement (chapter 9), he had to relocate his room to the third floor. In effect, this meant relinquishing a lab and teaching space that he had established as an appealing physical space to support science learning in a variety of ways. His lower-level room included spacious lab stations, convenient access to computers connected to the Internet, a classroom library, and easy access to whiteboards and tools such as the periodic table. As is clear from the accounts of Gale Seiler and Lacie Butler in chapter 3, resources such as these were not routinely distributed throughout the school, particularly since all science labs were located on the third floor. In fact, when Carambo first arrived at City High he invested considerable time and effort into creating a physical environment that included the resources he needed to enact a viable science education curriculum. In so doing, he built social capital with the SLC coordinator and the principal and was supported by them in his efforts to create the sort of program they wanted to see in the school. Through their assistance and his own energy and ingenuity, he was able to obtain the economic resources for equipment and supplies and thereby create an environment highly conducive to the teaching and learning of science, within what was formerly a metal workshop.

Carambo's transfer to the third floor was part of a school reorganization that reduced the number of SLCs to six. During his first year on the third floor, Carambo was a science teacher and had no formal administrative role, even though he was informally regarded by the principal as the leader of the science education group. Although there were large "double" labs, such as the one described in chapter 3, Carambo opted to have two adjacent classrooms as his home base for teaching science. The rooms were filthy and infested with rodents and litter thus, systematically, Carambo cleaned out the trash, disinfected the floors and walls, repainted the rooms and created a new classroom front that spanned the side walls of both previous classrooms. On the new front wall, he installed two large whiteboards and positioned a periodic table adjacent to one of them. Immovable lab stations were located at each end of this central space, allowing students to spread out when they were involved in lab work. For the second time at City High, Carambo transformed what was previously an unsuitable physical facility into a clean and spacious lab that could support the learning and doing of science in a variety of forms. In an important sense, he used his agency to adapt existing physical structures to support his vision of what was needed for a high-quality science education.

In relating this narrative about Carambo's efforts to create suitable physical spaces for science education, it is imperative that we emphasize that others helped him. Without the active support of the SLC coordinator and the principal, he would not have been able to secure sufficient physical space nor the economic resources to support his vision of creating spaces that were clean and functional; spaces that students felt good about learning in and did not send negative semiotic messages associated with teaching science, as did some of the other, less than adequate spaces on the third floor of City High (chapter 3).

In an analogous way, when Carambo became the SLC coordinator during the following year, he realized that the faculty in the SLC did not have a communal space in which to meet. Accordingly, just as he created a space for his own teaching of science, he transformed a room, that previously was a repository for old textbooks and broken equipment, into a faculty lounge. Once again, the school administration assisted him to acquire resources (i.e., suitable seating and a refrigerator) to create an environment that was inviting for faculty so that they could quietly prepare lessons, eat lunch, meet to discuss curricula, and otherwise interact formally and informally. Hence, Carambo was able to successfully access the physical structure and, through the use of social networks and his symbolic capital as a hardworking educational leader, he co-created a different form of physical structure that transformed the agency | structure dialectic, thereby affording the practices of other participants in the SLC.

Irrespective of the organizational arrangements in a school, teachers have agency to seek out materials and equipment to support the science curriculum. As Seiler and Butler demonstrated in chapter 3, the lack of particular equipment and materials does not necessarily mean that the only alternative is to focus learning on textbooks and copying notes from the chalkboard. Instead, Butler, through the use of resources on the Internet, identified different lessons to address many of the same and sometimes expanded goals. These lessons required the use of alternative materials that were more readily available and, as it happened, safer for students to use in the lab. Certainly, Butler and others have shown that teacher agency is not limited to seeking alternative lessons and can be afforded by speaking to a SLC coordinator, a vice principal, a professor from a nearby university, local business persons, or applying for a grant to obtain funding. For instance, while student teaching at City High, Jen Beers (chapters 6, 7, and 8) was able to obtain financial resources from within the school to purchase dissection specimens repeatedly requested by her students in the biology elective. When whole sheep heads (instead of brains removed from the cranium) arrived, she located assistance in cutting open the cranial bones at the University of Pennsylvania Veterinary School.

tices that reduce the time for teaching and learning then, having identified the contradiction collectively, a science faculty can act collectively to change practices and remove the contradiction to some extent. If teachers continue to act alone, or if they are expected to because of other within-school and district structures, it is unlikely that necessary transformations will occur and be sustained. In this regard, perhaps teachers can learn from their students who often demonstrate a greater sense of communalism and joint effort, of creating a posse or squad, and "having each other's back."

The Role of the School Administration

Clearly, from the examples we have encountered in this study, administrators can provide structural support in a variety of ways, mainly by assisting those who show the motivation to take a leadership role and catalyze improvements. However, not all means of support provided by administrators are successful. For example, when eliminating departments and inaugurating SLCs, the principal at City High imposed a rule system that created an informal organization that required science teachers to meet monthly to coordinate science education in the school. The reasons for the failure of this structural arrangement are probably associated with it being led by an assistant principal and mandated in conjunction with many other activities, and the existence of few incentives for science teachers to assume collective responsibility for the success of the meetings. Accordingly, relatively high levels of negative emotional energy, lack of solidarity, and determination on the part of most teachers to proceed in their own ways characterized most meetings. There was also a tendency for science teachers to not attend the meetings and, when attendance was required, the interactions among the faculty were not directed toward the creation of collective decisions and coresponsibilities for science education. Hence, rather than mandating attendance and imposing rules, it might have been preferable to create an organization in which interested science teachers could cogenerate collective agreements that could be presented to the school administrators, who would then act upon them.

In providing a discussion around the various ways teachers have exercised agency in relation to the physical structures of schooling, we emphasize here that structural changes do not come about through a simple equation in which teacher resolve equals improved classroom structure. That is, agency is not individual or collective but rather, both/and simultaneously. Without the larger administrative support system and without the dedication of teachers individually searching for new resources or for new ways to use current resources, structural changes (i.e., those that are physical or material in nature) cannot come about in any real sense.

Practices as Structures

Whereas urban schooling has been heavily criticized in terms of a lack of material resources, there is an equal tone of dissatisfaction regarding the human resources (i.e., teacher quality and, therefore, teaching practices). Thus, in this section, we discuss the practices of participants within a classroom as critical constituents of the structure of a field, in addition to material aspects. In the chapters in this volume, we learn from teacher and student past histories and their transformations into the present. In doing so, we glean important lessons about urban science education and possibilities for improvements into the future.

Within the classroom field, the practices of all participants comprise part of a structure to which agency is dialectically interwoven. From the perspective of the teacher, the critical question to ask is: What practices can structure the field in ways that will elicit appropriate forms of participation from all others? At the root of this question are teachers and students understanding and valuing one another's culture, accepting the cultural enactments of those who are culturally other, and interacting with them in successful ways that engender positive emotional energy. In chapter 2, the transformation of the structure of Kenneth Tobin's teaching is multilayered and catalyzed through changes in both the human resources (i.e., his interactions with the youth) and schematic aspects of structure (i.e., his valuing and understanding of the students' culture). For instance, the structure of Tobin's classroom altered as he came to see the students through lenses that were not tinted by the hegemonic ideology put forth by schools. He did not buy into the assumption (felt by many youth and expressed by many teachers and administrators) that his students were destined to failure. In addition, he learned to listen to student interactions without immediately perceiving that the tone and use of language was directed to nonscience talk. Before long, he realized that most of the student-student interactions were task-oriented and had the potential to contribute to science fluency.

Science as Culture

If science is to be viewed as cultural capital, it is important to draw attention to the practices and the schema that students are to learn, and since culture is enacted in fields, it is salient to consider where science is enacted and under what structural arrangements. Ideally, the science education learned in an urban high school will prove to be viable when it is enacted in other fields that are germane to the lives of students. Hence, science culture should be learned in such a way that it can be enacted as practices and schema in other fields and can afford the attainment of goals in those fields. Since much of culture is enacted in a field without awareness, it is important that students learn science in ways that allow them to enact it fluently when the structures create suitable resonances. Thus, the learning of science, as culture, extends beyond the learning of concepts and fixing "faulty" algorithms that students may bring with them to their science classes.

In this book, we have described the "doing" of science in terms of labs, discussions, interactive dialogues over contemporary science topics, reviews and syntheses of written materials in books and magazines, searches and syntheses of material obtained from the Internet, field trips to labs, businesses, and industries where science is practiced, and learning from videotapes and lectures. The key description is dynamic involvement in a variety of different activities so that the domain of science is sampled and students have chances to participate and learn by being involved with others who are also involved in science.

Learning Science as a Source of Agency

LaVan (chapter 14) undertook research in Martin's classroom and provided compelling evidence of the development of canonical science language and associated concepts from everyday talk as students interacted with one another in small groups and appropriated student-generated diagrams and everyday equipment to make sense of a common household problem of how to use a plunger to clear a blocked toilet. The microanalyses of students collaborating to solve problems in small groups provide insights into the production of science as culture and its fluent enactment as students enacted chains of successful interactions that generated positive emotional energy and solidarity among group members.

Various studies involving Carambo also provide insights into the development of science fluency among urban students. For example, presented in chapter 9, Carambo's research suggests that chains of successful interactions during the frog dissection produced significant cooperation within the lab group on which he focused, and there was ample evidence of the emergence of a foundation of solidarity and positive emotional energy on which future science education might build. When he allowed students to express and pursue their interests, many became deeply engaged, and they felt respected because he had chosen their ideas to include in the curriculum. Somewhat surprisingly, these students got involved and stayed involved for lengthy periods of time as they exhibited an array of practices in their doing of science. They were curious, persistent, patient, and resourceful as they dissected a frog and remained so, even after they had removed the heart (i.e., they then reached consensus on the need to dissect the heart). As necessary, these students accessed other resources to support their learning and reach their goals. As Carambo reports in chapter 9, this event represented a contradiction to the typical pattern of the students involved, who did not normally sustain their focus and participation in his science class. Although there is no evidence of subsequent successful science experiences for Jarvis, there is promise in learning from this one contradictory event, and the research surrounding this contradiction raises challenging questions for us to consider as we continue to advocate for improvements to urban science education.

The Centrality of Cultural, Social, and Symbolic Capital

To be an effective teacher in Papua New Guinea (PNG), Loman (chapter 15) had to adapt her practices in and out of the classroom to the practices and schema of her students and the community served by the school. It was not a quick process and yet the results of teaching in culturally adaptive ways were readily apparent. Her success in PNG, and subsequently in her teaching in different city schools, involved her willingness to earn the respect and trust of her students and to be regarded by them as fair. Loman showed that exchanges of social, cultural and symbolic capital were at the heart of becoming effective in all four school settings she described in her chapter. Also, she highlighted the importance of enacting curricula that were of potential use to her students in their lives out of school and that were not reliant on sophisticated equipment and expensive materials. Accordingly, she developed an approach that allowed to her to provide students with the direct experience of doing science, using simple equipment, even though school structures militated against materials-oriented curricula.

Martin's autobiography (chapter 13) traces her trajectory as a learner who was from home circumstances of poverty, raised by her single mother and caring grandparents, none of whom had attained high levels of education. In many regards, Martin's trajectory from high school to university, where she majored in science and pursued several graduate degrees, is highly unusual and against the more usual trend that applies to people with similar social and cultural histories. It is evident that Martin is talented and determined to succeed, qualities that served her well in her efforts to become an urban science teacher.

At Dewey Elementary, Martin struggled to earn the symbolic and social capital needed to be an effective teacher of inner-city youth. Gradually she learned how to interact with her students successfully, and thereby to build social networks and earn their respect and trust. The cultural capital produced while growing up in the rural South seemed to equip Martin to interact fluently with her urban students. In fact, toward the end of her year at Dewey Elementary, there were signs of her teaching being more fluent and of most students being cooperative and willing to learn from her. However, when Martin began to teach at Urban Magnet she realized that one size does not fit all. The teaching practices that were successful at Dewey Elementary were unsuccessful at Urban Magnet, and it became necessary for her to adapt her teaching to the structure of a very different type of urban school. Since the participants' practices are a critical component of the structure of a field, it was necessary for Martin to tailor her teaching practices to fit the dynamic and fluid structure of her science classes.

Science is a discourse that can be used to communicate with others and, as such, is a resource for building new social networks within and around science. Hence a goal of science education in urban high schools is to talk and do science in ways that enable social networks to be formed as resources for building a better social life. Specifically, by doing science, students will develop peer networks as well as learn how to interact with adults (usually their teacher). While a neglected focus has been on the necessity of students learning to build social networks that include adults, especially those from other class fields, we regard the creation of such social networks across gender and social borders as important outcomes of a science program. In effect, this means that the discourse of science can serve as a vehicle for social mobility. For example, learning how to interact with a teacher, who is older and likely from the middle class, provides students with important experiences in building social capital while crossing boundaries of gender and class. Moreover, learning how to create social networks with others, that include adult, middle-class, white teachers necessitates the building of practices and schema that have the potential to serve learners well in out-of-school contexts and in school subjects other than science. Hence the learning of science can be a source of agency because of the opportunities the process provides for learning how to build new forms of social capital.

The symbolic capital of learning science takes many forms. For many students, science is a subject that must be passed four times on the way toward the completion of a high school diploma. In that sense, all that is needed is a passing grade so that the necessary credits can be obtained to move successfully through high school. To others, science is seen as a prestigious subject in which it is difficult to do well. Accordingly, success in science is a marker of academic accomplishments that may serve as a ticket to other fields (e.g., entrance to a high school specialization or honors' studies; acceptance to college; consideration for certain forms of employment). If students are provided with opportunities to learn science in forms that are relevant and significant to everyday life, then there is an opportunity for them to designate their own symbolic markers on their participation and success in science.

Being informed about science in today's world, no doubt, is associated with identity in many ways as a person engages news accounts that include issues that are deeply science related, such as searching for life on Mars; capturing and analyzing stardust; understanding the potential threats to human life of Mad Cow Disease and Bird Flu; using DNA analyses in high-profile criminal investigations; fighting diseases such as SARS and AIDS and cancer; producing weapons of mass destruction; and using technology in the war on terrorism. Being able to use science to understand the unfolding news stories of the day can enhance the aesthetics of social life and assist students to see science as central in today's political and social world. However, enacting a curriculum that addresses issues such as those in the above list has implications for teachers who may feel constrained to follow heavily prescribed scope and sequence as urban school districts insist they enact standardized curricula and hold them accountable for the achievement of their students.

Student Interest–Driven Curriculum
A new biology curriculum at Charter High, developed by Seiler and enacted by Beers and another biology teacher at the school, represents an attempt to capture student interest and create resonances with the students' lives while addressing both the Pennsylvania life science standards and Charter High's academic literacy standards. It employs student inquiry groups similar to those enacted in Martin's class (chapter 14), and inquiry questions drive daily and weekly plans as well as each unit in the course. For example, students explore questions, such as, "Why is Yao Ming so tall?" and "Why is Verne Troyer (the actor, Mini-me) so small?" In doing so, they address the topics of genetic probability, types of inheritance, and gene expression, in addition to cell division, growth, and development. Inquiry groups foster science talk in class as students tackle content on a need-to-know basis. For example, understanding homeostasis and kidney function becomes critical to understanding why Alonzo Mourning had to quit the NBA. Combined with cogenerative dialogues and opportunities for student input and responsibility within the curriculum, such inquiry questions attempt to create positive emotional energy that can fuel engagement and learning of science in the ways that Carambo described in his classes and was found, at times, in the biology elective course (chapter 6 and chapter 7). Thus, we see the significance of science curriculum in which students learn meaningful content that is perceived as useful and relevant, and for teachers to systematically involve students in a review of the enacted science curriculum with a focus on involving them in ways that allow them to have voice as a source of social capital and opportunities

to build symbolic capital by gaining respect of teachers and school administrators.

Based on the sociology of emotions, there are grounds for spending some of each class period on activities that lead to success and the associated positive emotional energy of participating in successful interactions. If sufficient positive emotional energy can be associated with a science class then the probability increases that students will actively participate in science activities even if they are not aligned with their interests. The positive emotional energy of coparticipating with others, associated with successful interactions, can sustain a sense of group solidarity and create more positive emotional energy even for aspects of science that might not immediately appeal to students. This is an area that is in need of more research, but at this point we suggest that teachers enact some activities each day to address students' interests, thereby showing them respect, creating opportunities for successful interactions, and building positive emotional energy and solidarity.

Learning to Teach

Over the past six years, we have undertaken considerable research on learning to teach science in urban high schools. Experienced teachers, like Tobin (chapter 2) and Carambo (chapter 9) have initially struggled to teach in inner-city high schools, primarily because they were unable to read the culture of students, as it was enacted, and to adapt their own teaching to create fluency in the flow of interactions. Similarly, new teachers like Beers (chapters 7 and 8) have found it difficult to sustain productive learning environments. In all cases, the initial challenge appears to be for teachers to earn the symbolic capital of being regarded by students as *their* teacher. In order to accomplish this requisite goal, it is necessary to teach in ways that fit with the expectations of the students and show respect for them. One of the key ingredients to doing this is to identify the capital of students on which science learning can build. Elmesky's work has been illuminating of promising practices urban youth have developed in fields away from the classroom, that can serve as foundations for science learning, including *playin* (chapter 5) and argument (Elmesky, 2003). Likewise, Seiler's work has shown how cultural dispositions, such as the use of analogies and figurative speech generally acquired and practiced outside of school, can contribute to individual and collective learning in science class (chapter 6). However, while being *aware* of students' culture is important, perhaps more essential is actually *teaching* in ways that are adaptive to the students' practices and will create resonances for their dispositions which, when enacted, facilitate the learning of science.

In the case of Tobin, he learned about respect through his reading of the literature and adapted his teaching to the practices and dispositions of students by his teaching of them. By being and teaching *with* urban high school youth, Tobin learned how to teach in ways that got them involved and did not breach their participation. As he adapted his teaching to focus more on students who wanted to learn and to avoid interactions that produced negative emotional energy, his teaching became more fluent and was more in synchrony with the dispositions and associated practices of students. Accordingly, chains of synchronous interactions occurred and these produced successful outcomes and positive emotional energy. Over time, students appeared to enjoy being in the classroom to a greater extent than previously, and this allowed Tobin to create social networks and to earn symbolic capital, as *their* teacher. Ironically, as he earned that symbolic capital of teacher, he was able to be himself, and use more of his traditional ways of teaching.

In contrast to Tobin, who gradually acquired such dispositions by being *with* these students, Carambo already embodies dispositions similar to many of the students (i.e., verve, bodily movement, and expressive individualism). For instance, Carambo has a dynamic rhythm to his teaching that can create structures that resonate with many of the dispositions Boykin (1986) has identified. However, the sharing of dispositional practices was not enough for Carambo. In fact, even though he had been quite successful as an urban science teacher in Miami, Florida, Carambo struggled in his first semester at City High, largely because his practices were out of synchrony with those of the students and he had not earned the right to be *their* teacher (Roth, Tobin, Elmesky, Carambo, McKnight, and Beers, 2004). Just as Tobin had to find ways to build symbolic capital, so too was it necessary for Carambo to show that he was "cool" and thereby create social networks that allowed him to become a highly successful teacher from whom urban youth could learn a great deal about science and citizenry. Although we do not imply linearity, it seems critical for teachers to attend to earning the right to teach and other forms of symbolic capital—that is demonstrating practices that can earn respect, including oral fluency, athletic ability, performing raps, *playin*, and showing courage in adverse circumstances. Additionally, a significant part of teaching, demonstrated by Carambo, was enacting his knowledge of which practices to shut down and which to encourage. As he got to know his students better, he was able to teach more fluently and there was less evidence of shut downs. Interestingly, and consistent with Elmesky's chapter, he came to know when students were *playin* and increasingly showed his sense of humor and sense of the game by grinning and moving on with his teaching. When asked about his ability to *play*, one student expressed, "Oh, Mr. Carambo? He *play* all the time!"

On the basis of our studies of teaching and learning, it seems safe to say that interaction chains that produce negative emotional energy are best avoided. Often, teachers will have a sense of whether or not interactions are producing negative emotional energy and probably will not have to resort to microanalysis to identify which interactions lead to successful outcomes. When asynchronous actions occur, it might well be a sign that practices have to be adapted in order to find those that can produce higher levels of synchrony. Thus, the studies in this book suggest that science teachers might use emotional energy as a barometer of the extent to which teaching and learning are likely to produce successful outcomes.

Looking Forward while Looking Back

Science education, as praxis, is enacted in a field constrained by dialectical relationships between structure and agency (individual and collective), schema and practices (conscious and unconscious), and patterns of coherence and contradictions. Although this list is incomplete, it provides insights into the goal of enacting transformative science education in urban high schools. Conscious reviews of participation in a field (i.e., looking back) is a means of bringing salient practices of teachers and students to a level of consciousness, thus objectifying them in ways that afford research to identify patterns of coherence and associated contradictions. As we have shown in various chapters of the book, the research can then be discussed by stakeholders, including students, to permit the cogeneration of collective agreements intended to shape the path ahead, leading to improved learning environments. Hence, looking back fits recursively with planning for the future, realizing that when they are enacted, practices and associated schema restructure a field, thereby contributing to the agency of all participants. The result can be improved learning opportunities for all, especially if the looking back and looking forward are continuous and accompanied by commitments of individuals to act in the interests not only of their own personal learning, but with the success of all as a shared goal.

Creating Successful Chains of Interactions

Our studies of the teaching and learning of science in urban high schools illustrate the importance of enacting and sustaining successful chains of interactions in the science curriculum. If practices are oriented toward the creation of chains of successful interactions, then the vision of a transformative science education incorporates solidarity, with collective agreements and responsibilities as central conditions. Whether or not interactions are successful largely depends on exchanges of capital and a cycle in which capital is created and provides a foundation for building additional capital, possibly in different forms. Respect, a form of symbolic capital, seems to be at the base of this interactive spiral that is central to the learning of science. That is, unless the participants involved in interactions respect one another, there seems only a remote likelihood that successful interactions will occur in a sustaining way that can create positive emotional energy. If one of the interactants exhibits disrespect for the other, interactions can be asynchronous and chains of asynchronous interactions can develop, producing a buildup of negative emotional energy which can lead to dysfunctional learning environments. We regard it as a priority to ascertain which forms of interaction lead to the buildup of negative emotional energy so that enacted curricula can be tailored to avoid them. A key step in this process is engaging in coresearch with students so that science teachers can learn enough about their lives in order to show genuine interest in them and demonstrate authentic respect.

In addition to demonstrating respect, earning the respect of students is a significant priority for teachers and may take time, especially if efforts to interact initially produce negative emotional energy. Hence an a priori condition for successful teaching in urban high schools is to build respect for students and show them respect, thereby creating opportunities to establish social networks that become more expansive with time. Moreover, as teachers build social capital with students, it is essential that opportunities are provided for students to build social networks with the adults in the school, thus learning how to communicate effectively, to be respectful of the adults, and to earn their respect.

Students and Teachers Sharing Responsibility

We have shown in this research that teachers alone are not responsible and cannot be responsible for the quality of teaching. Therefore, it is not sufficient for teachers to focus only on what they must do to improve their teaching. As a form of culture, teaching is structured by a dynamic field, in which the curriculum is enacted and the agency of teachers is constituted in a dialectical relationship with that structure. Accordingly, effective teaching requires more than thorough planning and deep knowledge of science subject matter. Teaching can only be effective when students collectively support the goal for all participants to be successful. Thus individuals decide how to act, and when and how to appropriate resources in a field, but in so doing, they are aware of collective goals and collective responsibilities for the participation of all. To facilitate this awareness and associated practices, numerous studies have now been done to show how cogenerative dialogues with small groups and whole classes are fields that can produce cogenerated outcomes that can provide structures to dramatically improve the quality of science education.

Researchers at Charter High (chapter 8) and Urban Magnet (chapter 16) were successful in obtaining students' perspectives on what is salient and then discussing with them how the curriculum might be transformed so as to produce more favorable outcomes. To assist students to assist us, the university and teacher researchers taught their students about social and cultural theory, including the forms of capital, agency, structure, solidarity, and emotional energy. They were then able to search through videotapes of their science classes and identify vignettes that were salient to the quality of teaching and learning. Then, in cogenerative dialogues, conversations occurred about the patterns of coherence and associated contradictions. Dealing with the contradictions became a catalyst for changing the learning environments and especially renegotiating the roles of teachers and students. Scheduling regular cogenerative dialogues with small groups that include different types of students, and regularly (but less frequently) to have whole-class cogenerative dialogues appeals as a highly productive activity. Collective agreements on what is to be accomplished, the tools to be used and how to use them, the rules to apply in the classroom and new roles to be enacted by participants are worthwhile foci for cogenerative dialogue. When these are well conducted, there are significant cycles in which social, cultural, and symbolic capital are developed by all participants—capital that can potentially be enacted, with support from the group, as science is taught and learned. Hence, the agency afforded by participation in cogenerative dialogues has the potential to transfer into the field of science education for the benefit of all participants.

The Role of Out-of-School Fields

Since many urban teachers are culturally, socially, and historically different than their students, it is important that the teachers carefully study the social lives of their students and that the students learn about the social lives of their teachers.

This represents an extended project and might be beyond the resources of most teachers, who have full-time jobs and busy lives away from school. Accordingly, we regard it as a priority to undertake intensive studies of the different out-of-school fields in which urban youth are successful and to share these understandings with others. In the study of capital produced in other fields, we explore interactions and events using both mesoscopic and microscopic lenses. When practices are identified as constituents of successful interaction chains, we should search for their applications in other fields. Just as we have done with arguing, *playin*, and figurative speech, so too must we search for cultural capital from other fields that might serve as a foundation for the building of science fluency.

Although we are pleased with the success we have had in linking the dispositions identified by Boykin to the learning of science, we are determined to add to those dispositions, taking care not to stereotype along boundaries defined by race, gender, and class. For example, we find it important to emphasize that, while Boykin identifies dispositions as rooted in African culture, our research has shown that the emergence of particular ways of being are shaped by the structures of the fields in which the dispositions develop, rather than by ethnicity and race. That is, although others may read our research and attempt to link the identification of a disposition of argument or *playin* to pertain exclusively to African American and/or black individuals, these dispositions are shaped by the structures of fields in which youth participate. Thus patterned actions or common practices found within or across urban schools in a city like Philadelphia are not indicative of a culture of a particular ethnic or racial group of individuals, but rather evidence of similar structural components of the social fields from which the school populations emerge. This is a further indication of common material and human resources and rule systems within the fields from which youth are coming such that, more than sharing a common skin tone or ancestry, they share realms of experiences (i.e., living in tightly knit, all-encompassing families) in neighborhoods, similarly structured by inequitable societal phenomena such as institutional racism, poverty, or partiality to Eurocentric masculine values. This recognition of the centrality of structure in shaping (and being shaped by) dispositions is essential for forming understandings that these dispositions can and are embodied by a large range of individuals (who may or may not hold similar goals). As many of the teachers studied in this book have shown so convincingly, if classroom and school fields can be structured to resonate with the dispositions of the participants and if the dispositions are not truncated when the resonances occur, there is a high likelihood that cultural enactment will flow and opportunities for learning will abound.

Optimizing the Flow

Creating the conditions to optimize flow, with the goal of creating culture that can be enacted fluently in multiple fields, is a raison d'être for science. We see three prongs to an approach that is likely to succeed. First, through the use of cogenerative dialogues that incorporate micro and meso analyses of selected video vignettes, teachers and students can become conscious of practices about which they were not fully aware and can identify structures to optimize participation of desired forms. Second, by teaching with the students intensively and extensively, the practices of participants are likely to become mutually adaptive, especially if all participants are aware of and support a goal of minimizing interactions associated with negative emotional energy and maximizing those that are associated with positive emotional energy. Third, whereas teachers and students can learn about different approaches to teaching and learning through the first two prongs of the approach, there is significant value assigned to experiencing different teaching and learning contexts in the same school. Of course, students experience multiple teachers in a day and teachers usually are responsible for multiple classes a day. Even so, we regard it as a valuable professional development tool for teachers to coteach science classes on a regular basis. Our research on coteaching is reported in several parts of this book and is dealt with intensively elsewhere (e.g., Roth and Tobin, 2002). We exhort teachers to coteach with a variety of others, including school administrators, local university personnel, colleagues, students from advanced classes and student teachers. The purposes of coteaching are to experience forms of teaching that are different and to learn from them, not only at the conscious level through cogenerative dialogues and micro and meso analyses, but also by simply teaching with other teachers. In the process, by teaching at another's elbow, new forms of capital and agency can be built due to the different structural arrangements that arise in the presence of the practices of other teachers.

Will the World Be Waitin'?

The research in this book transcends micro, meso, and macro levels of social life. We have struggled intellectually to interconnect these different levels of social life with powerful social theorems such as the agency | structure dialectic. We endeavor to build deep understandings of the interactions that unfold as participants engage resources in the dynamic field of science education and subsequently enact culture in myriad social fields in and out of school. We now have the intellectual lenses to explore the transformative potential of science education macroscopically and to identify patterns of coherence and contradictions which can then be investigated within multiple fields by investigating salient events at meso and micro levels. Additionally, we see the advantages of examining actions, practices, interaction chains, and the emergence of solidarity across all three levels, and examining the patterns of coherence in relation to the contradictions, which will always exist as long as patterns occur. Making sense of contradictions and patterns of coherence, across micro, meso, and macro boundaries is a goal that is at the heart of whether or not students benefit from science education: whether the world be waitin' for them. Integral to this possibility is that science education can create the potential for agency that can transcend cultural, social, political, and economic boundaries.

Studies that examine social life across these levels are essential, if we are to seriously address the goal of transformative science education, especially if the research is intensive and extensive and fully employs researchers from different niches in the fields of study. The research will inevitably involve many types of participant researchers from outside classrooms (e.g., guardians, family members, coaches, coworkers, and religious leaders) since we will necessarily include fields of study such as home, work, recreation, and religion. Our interest extends beyond the flow of culture from the science classroom to other fields, to include the flow of science culture from other fields to the classroom. As culture is produced and enacted in fields, the identities of learners change because of their learning and expanded agency; however, science will also be transformed as it is produced and reproduced by urban youth. If the endpoints of science education will be characteristic of urban youth and build on extant culture, it only makes sense, in the design of science curricula, to fully involve youth as coresearchers, teacher educators, and curriculum designers. In this way, the starting points are then reflective of the perspectives and life histories of urban youth and increase the likelihood that the outcomes of science education (practices and associated schema) will reflect the uniqueness of the social, cultural, and historical trajectories of those who will become fluent in science and use it macroscopically to catalyze ascending trajectories through social space.

Bianchini, J. A., L. M. Cavazos, and J. V. Helms. (2000). From professional lives to practice: Science teachers and scientists' views of gender and ethnicity in science education. *Journal of Research in Science Teaching* 37: 511-547.

Boden, D. (1994). *The business of talk: Organization in action.* Cambridge, UK: Polity Press.

Bourdieu, P. (1977). Cultural reproduction and social reproduction. In *Power and ideology in education*, ed. J. Karabel and A. H. Halsey, 487-511. New York: Oxford University Press.

———. (1986). The forms of capital. In *Handbook of theory and research for the sociology of education*, ed. J. G. Richardson, 241-258. New York: Greenwood Press.

———. (1992). The practice of reflexive sociology (The Paris workshop). In *An invitation to reflexive sociology*, ed. P. Bourdieu and L. J. D. Wacquant, 216-260. Chicago, IL: The University of Chicago Press.

Bourdieu P., and L. Wacquant. (1992). *An invitation to reflexive sociology.* Chicago, IL: The University of Chicago Press.

Boykin, A. W. (1986). The triple quandary and the schooling of Afro-American children. In *The school achievement of minority children: New perspectives*, ed. U. Neisser, 57-92. Hillsdale, NJ: Erlbaum.

Brett, J., and L. K. Stroh. (2003). Working 61 plus hours per week: Why do managers do it? *Journal of Applied Psychology* 88: 67-78.

Brown, A. (1992). Design experiments: Theoretical and methodological challenges in creating complex interventions in classroom settings. *The Journal of the Learning Sciences* 2: 141-178.

Burgess, N. (1994). Gender roles revisited: The development of "women's place" among African American women in the United States. *Journal of Black Studies* 29: 391-401.

Chiapetta, E. L. (2001). Build conceptual knowledge before laboratory work. In *Cases in middle and secondary science education: The promise and dilemmas*, ed. T. R. Koballa and D. J. Tippins, 140-141. Upper Saddle River, NJ: Merrill.

Cochran-Smith, M., and S. L. Lytle. (1990). Research on teaching and teacher research: The issues that divide. *Educational Researcher* 19: 2-11.

Cole, M., and Y. Engeström. (1993). A cultural-historical approach to distributed cognition. In *Distributed cognitions: Psychological and educational considerations*, ed. G. Salomon, 1-46. New York: Cambridge University Press.

Collins, P. H. (1990). *Black feminist thought: Knowledge, consciousness, and the politics of empowerment.* New York: Routledge.

———. (1998). *Fighting words: Black women and the search for justice.* Minneapolis: University of Minnesota Press.

———. (2000). *Black feminist thought: Knowledge, consciousness, and the politics of empowerment.* New York: Routledge.

Collins, R. (1993). Emotional energy as the common denominator of rational choice. *Rationality and Society* 5: 203-230.

———. (2004). *Interaction ritual chains.* Princeton, NJ: Princeton University Press.

Csikszentmihalyi, M. (1990). *Flow: The psychology of optimal experience.* New York: Harper Perennial.

Darling-Hammond, L. (1999). Reforming teacher preparation and licensing: Debating the evidence. *Teachers College Record* 102: 28-56.

Darling-Hammond, L., and B. Falk. (1997). *Using standards and assessments to support student learning: Alternatives to grade retention.* New York: National Center for Restructuring Education, Schools, and Teaching.

Delpit, L. (1988). The silenced dialogue: Power and pedagogy in educating other people's children. *Harvard Educational Review* 58: 280-298.

———. (1993). *Other people's children: Cultural conflict in the classroom.* New York: The New Press.

———. (2002). No kinda sense. In *The skin we speak,* ed. L. Delpit, and J. K. Dowdy, 31-38. New York: The New Press.

Dewey, J. (1916). *Democracy and education: An introduction to the philosophy of education.* New York: The Macmillan Company.

———. (1944). *Democracy and education.* New York: Free Press.

Dinkelman, T. (2000). An inquiry into the development of critical reflection in secondary student teachers. *Teaching and Teacher Education* 16: 195-222.

Donnelly, J. F. (1999). Schooling Heidegger: On being in teaching. *Teaching and Teacher Education* 15: 933-949.

Eckert, P. (1989). *Jocks and burnouts: Social categories and identity in the high school.* New York: Teachers College Press

Eldon, M., and M. Levin. (1991). Cogenerative learning: Bringing participation into action research. In *Participative action research,* ed. W. F. Whyte, 127-142. Newbury Park, CA: Sage.

Elmesky, R. (2001). *Struggles of agency and structure as cultural worlds collide as urban African American youth learn physics.* Unpublished doctoral dissertation, The Florida State University, Tallahassee.

———. (2003). Crossfire on the streets and into the classroom: Meso/micro understandings of weak cultural boundaries, strategies of action and a sense of the game in an inner-city chemistry classroom. *Cybernetics and Human Knowing* 10: 29-50.

Elmesky, R., and K. Tobin. (2003). *Expanding our understandings of urban science education by expanding the roles of students as researchers.* Manuscript submitted for publication.

Engeström, Y. (1987). *Learning by expanding: An activity-theoretical approach to developmental research.* Helsinki: Orienta-Konsultit.

———. (1999). Activity theory and individual and social transformation. In *Perspectives on activity theory,* ed. Y. Engeström, R. Miettinen, and R. L. Punamaki, 282-297. New York: Cambridge University Press.

Explore IT. (2002). Retrieved January 1, 2003, from www2.mtroyal.ab.ca/~tnickle/ExploreIT

Fine, M. (1994). *Chartering urban school reform: Reflections on public high schools in the midst of change.* New York: Teachers College Press.

Fordham, S. (1996). *Blacked out: Dilemmas of race, identity and success at Capital High School.* Chicago: The University of Chicago Press.

Freire, P. (1970). *Pedagogy of the oppressed.* New York: Continuum.

———. (1993). *Pedagogy of the city.* New York: Continuum.

Gabel, D. (1999). Improving teaching and learning through chemistry education research: A look to the future. *Journal of Chemical Education* 76: 548-553.

Geuther, R., and H. Olmstead. (1996). Diaper Derby. Dow/NSTA Summer Workshop Lesson Plan. Retrieved April 25, 2002, from http://thechalkboard.com/Corporations/Dow/Programs/NSTA_Lessons/Diapers.html

Guba, E., and Y. Lincoln. (1989). *Fourth generation evaluation.* Newbury Park, CA: Sage.

Haberman, M. (1991). The pedagogy of poverty versus good teaching. *Phi Delta Kappan* 73: 290-294.

Hogle, L. F. (1995). Standardization across domains: The case of organ procurement. *Science, Technology, and Human Values* 20: 482-500.

Hogan, K. (1999). Depth of sociocognitive processing in peer groups' science discussions. *Research in Science Education* 29: 457-477.

hooks, b. (1995). *Killing rage: Ending racism.* New York: Henry Holt and Company.

Hurtado, A. (1996). *The color of privilege: Three blasphemies on race and feminism.* Ann Arbor: University of Michigan Press.

Ingersoll, R. (1997). Teacher turnover and teacher quality: The recurring myth of teacher shortages. *Teachers College Record* 99: 41-44.

James, J., and T. Sharpley-Whiting. (2000). *The black feminist reader.* Malden, MA: Blackwell.

Johnstone, A. H. (1982). Macro- and microchemistry. *School Science Review* 64: 377-379.

Kavanagh, D., and L. Araujo. (1995). Chronigami: Folding and unfolding time. *Accounting, Management and Information Technologies* 5: 103-121.

Kincheloe, J. L. (1998). Critical research in science education. In *International handbook of science education*, ed. B. J. Fraser and K. G. Tobin, 1191-1205. Dordrecht, Netherlands: Kluwer.

Kozol, J. (1991). *Savage inequalities: Children in America's schools.* New York: Crown.

———. (1995). *Amazing grace: The lives of children and the conscience of a nation.* New York: Crown.

Laird, S. (2002). Befriending girls as an educational life-practice. In *Philosophy of Education*, ed. S. Fletcher, 73-81. Urbana, IL: Philosophy of Education Society.

Lamont, M., and A. Lareau. (1998). Cultural capital: Allusions, gaps and glissandos in recent theoretical developments. *Sociological Theory* 6: 153-168.

Lather, P. (1986). Research as praxis. *Harvard Educational Review* 56: 257-277.

Lave, J. and E. Wenger. (1991). *Situated learning: Legitimate peripheral participation.* New York: Cambridge University Press.

Lee, C. D. (1992). Literacy, cultural diversity, and instruction. *Education and Urban Society* 24: 279-291.

———. (2001). Is October Brown Chinese? A cultural modeling activity system for underachieving students. *American Educational Research Journal* 38: 97-114.

Lemke, J. L. (1990). *Talking Science: Language, learning and values.* Norwood, NJ: Ablex Publishing Company.

———. (2000). Across the scales of time: Artifacts, activities, and meanings in ecosocial systems. *Mind, Culture, and Activity* 7: 273-290.

Lindbeck, A., and D. J. Snower. (2000). Multi-task learning and the reorganization of work: From Tayloristic to holistic organization. *Journal of Labor Economics* 18: 353-376.

Lytle, J. (April, 1998). *Using chaos and complexity theory to inform high school redirection.* Paper presented at the meeting of the American Educational Research Association, San Diego, CA.

MacLeod, J. (1995). *Ain't no makin' it: Aspirations and attainment in a low-income neighborhood.* Boulder, CO: Westview Press.

Martin, S. (2002, March). *Not so strange in a strange land: An autobiographical approach to becoming a science teacher in an urban high school.* Paper presented at the 23rd Annual Ethnography in Education Research Forum, University of Pennsylvania, Philadelphia, PA.

Material safety data sheet for chloroform, number C2915. (2001). Mallinckrodt Chemicals. Retrieved October 2002, from www.msdssearch.com

Mehan, H. (1993). Beneath the skin and between the ears: A case study in the politics of representation. In *Understanding practice: Perspectives on activity and context*, ed. S. Chaiklin and J. Lave, 241-268. New York: Cambridge University Press.

Mirza, H. S. (1992). *Young, female and black.* New York: Routledge.

Moi, T. (1999). *What is a woman?* New York: Oxford University Press.

Murrell, P. (2001). *The community teacher.* New York: Teachers College Press.

National Research Council. (1996). *National science education standards.* Washington, DC: National Academy Press.

Olitsky, S. (2003). *Balancing equations as collective activity: Interaction ritual, teacher-student roles, and the pursuit of a community of practice.* Unpublished manuscript, University of Pennsylvania, Philadelphia, PA.

Opper, M., and L. Spencer. (1997). Research and Development of a Polymer. Dow/ NSTA Summer Workshop Lesson Plan. Retrieved April 25, 2002, from http://the chalkboard.com/Corportions/Dow/Programs/97NSTA_Lessons/Lessons /unit6.html

Orna, M. V., J. O. Schreck, and H. Heikkinen, ed. (1998). *ChemSource Sourcebook.* Philadelphia, PA: Chemical Heritage Foundation.

Polman, J. L., and R. D. Pea. (2001). Transformative communication as a cultural tool for guiding inquiry science. *Science Education* 85: 223-238.

Ricoeur, P. (1991). *From text to action: Essays in hermeneutics, II.* Evanston, IL: Northwestern University Press.

Rodgers, C. (2002). Defining reflection: Another look at John Dewey and reflective thinking. *Teachers College Record* 104: 842-866.

Roman, L., and M. Apple. (1990). Is naturalism a move away from positivism? Materialist and feminist approaches to subjectivity in ethnographic research. In *Qualitative inquiry in education: The continuing debate*, ed. E. Eisner and A. Peshkin, 38-73. New York: Teachers College Press.

Roth, W.-M. (2001). Situating cognition. *The Journal of the Learning Sciences* 10: 27-61.

———. (2002). *Being and becoming in the classroom*. Westport, CT: Ablex/Greenwood.

Roth, W.-M., D. V. Lawless, and D. Masciotra. (2001). Spielraum and teaching. *Curriculum Inquiry* 31: 183-207.

Roth, W.-M., M. K. McGinn, C. Woszczyna, and S. Boutonné. (1999). Differential participation during science conversations: The interaction of focal artifacts, social configuration, and physical arrangements. *The Journal of the Learning Sciences* 8: 293-347.

Roth W.-M., and K. Tobin. (2001). Learning to teach science as praxis. *Teaching and Teacher Education* 17: 741-762.

———. (2002). *At the elbows of another: Learning to teach through coteaching*. New York: Peter Lang.

Roth, W.-M., K. Tobin, R. Elmesky, C. Carambo, Y. McKnight, and J. Beers. (2004). Re/making identities in the praxis of urban schooling: A cultural historical perspective. *Mind, Culture, and Activity* 11: 48-69.

Roth, W.-M., K. Tobin, and S. Ritchie. (2003). *Talk about the right time! An organizational perspective on teaching and learning science*. Manuscript submitted for publication.

Roth, W.-M., K. Tobin, A. Zimmermann, N. Bryant, and C. Davis. (2002). Lessons on and from the dihybrid cross: An activity-theoretical study of learning in coteaching. *Journal of Research in Science Teaching* 39: 253-282.

Roth, W.-M., and M. Welzel. (2001). From activity to gestures and scientific language. *Journal of Research in Science Teaching* 38: 103-136.

Schiffrin, D. (1996). Narrative as self-portrait: Sociolinguistic constructions of identity. *Language in Society* 25: 167-203.

Schön, D. A. (1985). *The design studio: An exploration of its traditions and potentials*. London: RIBA Publications for RIBA Building Industry Trust.

Science Education for Public Understanding Program (SEPUP). (1992). *Plastics in our lives*. Ronkonkoma, NY: Lab Aids.

Seiler, G. (2001). Reversing the "standard" direction: Science emerging from the lives of African American students. *Journal of Research in Science Teaching* 38: 1000-1014.

———. (2002). *Understanding social reproduction: The recursive nature of coherence and contradiction within a science class*. Unpublished doctoral dissertation, University of Pennsylvania, Philadelphia.

Seiler, G., K. Tobin, and J. Sokolic. (2001). Design, technology, and science: Sites for learning, resistance, and social reproduction in urban schools. *Journal of Research in Science Teaching* 38: 746-767.

Sewell, W. H. (1992). A theory of structure: Duality, agency and transformation. *American Journal of Sociology* 98: 1-29.

———. (1999). The concept(s) of culture. In *Beyond the cultural turn,* ed. V. E. Bonell and L. Hunt, 35-61. Berkeley: University of California Press.

Shapiro, B. (2000). Creating objects of meaning in science. In *Cases in middle and secondary science education: The promise and dilemmas,* ed. T. R. Koballa and D. J. Tippins, 97-109. Upper Saddle River, NJ: Merrill.

Smith, E. (2002). Ebonics: A case history. In *The Skin We Speak,* ed. L. Delpit and J. K. Dowdy, 15-30. New York: The New Press.

Suchman, L. A., R. Pea, J. S. Brown, and C. Heath. (1987). *Plans and situated actions: The problem of human-machine communication.* New York: Cambridge University Press.

Swidler, A. (1986). Culture in action: Symbols and strategies. *American Sociological Review* 51: 273-286.

Tobin, K. (1998). Issues and trends in the teaching of science. In *International handbook of science* education, ed. B. J. Fraser and K. Tobin, 129-151. Dordrecht, Netherlands: Kluwer.

———. (2000). Becoming an urban science educator. *Research in Science Education* 30: 89-106.

———. (2002). Beyond the bold rhetoric of reform: (Re)Learning to teach science appropriately. In *Science education as/for sociopolitical action,* ed. W-M. Roth and J. Desautels, 125-150. New York: Peter Lang.

Tobin, K., G. Seiler, and E. Walls. (1999). Reproduction of social class in the teaching and learning of science in urban high schools. *Research in Science Education* 29: 171-187.

Tyree, O. (1993). *Flyy girl.* New York: Simon and Schuster.

Utt, R. D. (2000). Cities and suburbs: Promoting innovative solutions to community problems. Retrieved October, 2001, from www.heritage.org/issues/chap13.html

Walby, S. (1997). *Gender transformations.* London: Routledge.

———. (2000). Analyzing social inequality in the twenty-first century: Globalization and modernity restructure inequality. *Contemporary Sociology—A Journal of Reviews* 29: 813-818.

Weiler, J. (2000). *Codes and contradictions: Race, gender, identity, and schooling.* Albany, NY: SUNY Press.

Wertsch, J. V. (1991). *Voices of the mind: A sociocultural approach to mediated action.* Cambridge, MA: Harvard University Press.

———. (1998). *Mind as action.* New York: Oxford University Press.

Whitman, G. (2000). Neighborhood schools prepare few for college. *Philadelphia Public School Notebook* 7: 1 and 9.

Wilson, W. J. (1987). *The truly disadvantaged: The inner city, the underclass and public policy.* Chicago, IL: The University of Chicago Press.

———. (1996). *When work disappears: The world of the new urban poor.* New York: Knopf.

————. (1999a). Affirming opportunity. *American Prospect* 46: 61-64.

————. (1999b). When work disappears: New implications for race and urban poverty in the global economy. *Ethnic and Racial Studies* 22: 479-499.

Wong, E. D. (1993). Understanding the generative capacity of analogies as a tool for explanation. *Journal of Research in Science Teaching* 30: 1259-1272.

Yerrick, R. K., E. Doster, J. S. Nugent, H. M. Parke, and F. E. Crawley. (2003). Social interaction and the use of analogy: An analysis of preservice teachers' talk during physics inquiry lessons. *Journal of Research in Science Teaching* 40: 443-463.

About the Contributors

Jennifer Beers has been teaching biology and general science at Urban Charter for three years. She has also been a mentor teacher for students from the teacher education program at the University of Pennsylvania. Jennifer holds a bachelor's degree in biology and master's degree in education from the University of Pennsylvania. Her research primarily focuses on the impact of cogenerative dialogues on school structures, teacher praxis, and student identity as science learners.

Lacie Butler is currently a first-year teacher of biology and earth and space science at Abington Junior High School. While earning her master's degree in education from the University of Pennsylvania, she taught biological sciences for a year at City High in Philadelphia, undertaking research on the teaching and learning of science in urban high schools for her thesis. Her continued research interest is on the impact of material and human resource application and appropriation on the structure and agency experienced by students and teachers.

Cristobal Carambo has been teaching science for eight years. He is a teacher of chemistry and physical science at City High School where he is the academic leader for the Science Engineering and Mathematics Academy. He is also a mentor for student teachers from the University of Pennsylvania's graduate school of education. He holds a master's degree from Florida State University. Cristobal is currently enrolled in the Master's of Chemistry Education program at the University of Pennsylvania and he is also pursuing a doctorate in science education at Curtin University. His current research focus is on the interrelationship between student agency, inquiry learning, and scientific literacy in the science classroom.

Rowhea Elmesky is an assistant professor at Washington University where she continues a program of research on the teaching and learning of science in urban high schools. Rowhea has her undergraduate degree in elementary education and graduate degrees in science education, including her doctorate, from Florida State University. Her main contributions to the science education field have been in developing macro, meso, and micro level understandings regarding:
1) the ways in which practices, resources, and schemas from social fields outside of the classroom shape what occurs within;
2) the identification of students' cultural capital, including conscious and unconscious practices; and
3) the development of scientific identity and science "fluency."

Sarah-Kate LaVan is an assistant professor at Temple University. She recently completed a master's degree in chemistry education and a doctoral degree in science education at the University of Pennsylvania. Sarah-Kate has taught science in various urban and suburban settings in Philadelphia and Miami. Her research interests involve examining the development of classroom communities and structures that foster equity, agency, and collective responsibility. A special emphasis in this book is her research on the use of cogenerative dialogue, video analysis, and social theory as means to expand the roles of teachers and students in the classroom.

Linda Loman has been teaching in a variety of settings for the last seven years. She is currently a teacher of physics and physical science at Glenwood Springs High School in Colorado, where she also coaches basketball and soccer. Linda began her teaching career in the Peace Corps and then earned her master's of education degree from Temple University while teaching in Philadelphia. She is currently pursuing a doctorate in science education at Curtin University. Her research focus is on her teaching practices when teaching out of field and examining the benefits of working with student researchers.

Sonya Martin, after graduating from the University of Pennsylvania with a master's degree in elementary education, taught science at both the elementary and high school levels within the city of Philadelphia. Drawing on her master's degree in chemistry (also from the University of Pennsylvania) and her teaching experiences she is currently researching the larger questions surrounding student-teacher relations, specifically student and teacher identity and how identity can influence and be influenced by the culture of science. Sonya completed her doctorate in science education at Curtin University and is an assistant professor in science education at Queens College in New York City.

Catherine Milne is an assistant professor in science education at New York University where she conducts research in urban science classrooms and teaches a range of courses in science education to graduates and undergraduates seeking to become science teachers in middle and high schools. Catherine taught high school science for many years in Australia before coming to the University of Pennsylvania as a post doctoral research fellow to conduct research associated with the teaching and learning of chemistry at college and high school levels. Her research interests include the storied nature of science, the role of representations in the culture of science, and the nature of effective science learning and teaching in urban contexts.

Stacy Olitsky is a graduate student working toward a doctorate in education and a master's degree in sociology at the University of Pennsylvania, where her research focuses on the teaching and learning of science in urban secondary schools. Currently, she is conducting an ethnographic study of an eighth grade science classroom, where she has been investigating differences between in-field and out-of-field science teaching, activity theory as a model for local change, the impact of school choice on students' identities, the role of social capital and cultural capital in the classroom, and interaction ritual and the formation of solidarity surrounding science learning.

Tracey Otieno served as a scientist and engineer in the United States, Ghana, and Kenya for eight years. She became a science teacher and taught chemistry and physics in Philadelphia High schools for four years. Tracey's chapter focuses on her final year of high school teaching at a neighborhood high school in Philadelphia. For the past two years Tracey has been a research associate at the University of Pennsylvania, where she undertook research on the teaching and learning of chemistry at college and high school levels.

Wolff-Michael Roth is Lansdowne Professor of applied cognitive science at the University of Victoria. His research focuses on knowing and learning in science and mathematics, which he studies in formal schooling, professional practice, and in everyday life situations. Being interdisciplinary, his articles are published in different disciplines, including sociology of science, linguistics, and areas in education (mathematics, science, teaching, curriculum studies).

Kathryn Scantlebury is an associate professor in the department of chemistry and biochemistry at the University of Delaware. As a science teacher educator, Kate coordinates the secondary science education program and undertakes research on gender equity and social justice in science education. Her recent activities in urban schools have extended her research on equity to include factors of race, ethnicity, and social class as factors associated with the learning of science.

Gale Seiler is an assistant professor at the University of Maryland, Baltimore County where she teaches and supervises student teachers. She was a high school science teacher for sixteen years, teaching culturally diverse students in a variety of settings from South America to Baltimore. Her research examines how curricula and classrooms can be restructured to build on the cultural practices and dispositions of urban, African American students and the culturally specific ways in which African American students participate in school science. In addition, she is interested in the preparation of teachers both in and for urban schools, and collaborative research involving multiple participants. Chapter 6 in this volume is derived from her Ph.D. dissertation at the University of Pennsylvania.

Melissa Sterba is an assistant professor at Temple University where she teaches educational law and social foundations of education. While a doctoral student at the University of Pennsylvania she studied gender equity issues in mathematics and science classes in urban high schools. Her focus on sexuality and violence of female youth from inner-city schools is grounded in extensive professional experience as a lawyer involved with the intersection of minority youth and the criminal justice system.

Kenneth Tobin is a Presidential Professor in The Graduate Center at The City University of New York. Prior to commencing a career as a teacher educator, Tobin taught high school science and mathematics in Australia and was involved in curriculum design. His research interests are focused on the teaching and learning of science in urban schools, which involve mainly African American students living in conditions of poverty. A parallel program of research focuses on coteaching as a way of learning to teach in urban high schools.

Index

Abraham, Anita, 59, 103, 304–5
access codes, 2
activity theory, xxi, xxiii, 170;
advantages of, 27; application of,
283–85; change promoted by, 284;
emergence of, 25; exploration of,
26; to review classroom practices,
178. *See also* Cultural Historical
Activity Theory (CHAT)
African American students, xx; class
engagement among, 115;
contradictions experienced by, 86;
dispositional resources of, 91;
earning respect of, 35–36; gate-
keeping's influence on, 127;
inadequate education's influence
on, 2; life experiences of, 113, 125;
oppression of, 9; oral tradition of,
xx–xxi; practices demeaning to, 68;
practices/dispositions of, 114;
school underrepresentation of, 3;
science learning of, 129; shut-down
practices' influence on, 30;
solidarity among, 29; Standard
English spoken by, 120; in
synchrony with, 36; talents of, 40;
teacher expectations of, 119, 135;
teaching of, 8; underevaluation of,
127; at Urban Magnet, 6; verbal

abilities of, 121. *See also* Black
English
African American women: family
attitude toward, 197; lifeworlds of,
204; motherhood, symbol for, 199;
racism/sexism subjecting, 197; in
school, 200–204; self-definition of,
201; social capital opportunities for,
205; stereotyping of, 182
African-rooted culture, 91
agency: challenges to, 50;
contributing factors to, 51; cultural
capital impacted by, 201, 203;
exhibited in classroom, 158; female
sexuality linked to, 182–83;
through ghetto analogies, 123;
of girls, 204–5; human resources
creating, 58; improved through
videotaping, 162; language as
source of, 256; learning linked to,
119; meaning of, 25; at micro level,
25; obstacles overcome with, 63;
playin's role in, 91, 102; power of
collective, 305–6; redistribution of,
155; renewal of, 53; in science
class, 200, 299; social capital
impacted by, 201, 203; as social
construction, 305; in social life, 88;
structure linked to, 169; student